Process Economics

James R. Couper, D.Sc. *Professor Emeritus, The Ralph E. Martin Department of Chemical Engineering, University of Arkansas—Fayetteville (Section Editor)*

Darryl W. Hertz, B.S. *Manager, Front-End Loading and Value-Improving Practices Group, KBR (Front-End Loading, Value-Improving Practices)*

(Francis) Lee Smith, Ph.D., M.Eng. *Principal, Wilcrest Consulting Associates, Houston, Texas (Front-End Loading, Value-Improving Practices)*

GENERAL COMMENTS

ACCOUNTING AND FINANCIAL CONSIDERATIONS

CAPITAL COST ESTIMATION

MANUFACTURING-OPERATING EXPENSES

FACTORS THAT AFFECT PROFITABILITY

PROFITABILITY

OTHER ECONOMIC TOPICS

CAPITAL PROJECT EXECUTION AND ANALYSIS

GLOSSARY

Nomenclature and Units

Symbol	Definition	Units
A_1	Annual conversion expense at production rate 1	\$
A_{TC}	Annual capital outlay	\$
B	Constant	Dimensionless
C	Cost of equipment	\$
C_B	Base cost of carbon steel exchanger	\$
CE	*Chemical Engineering* cost index	Dimensionless
$(C_{FC})_{BL}$	Battery-limits fixed capital investment	\$
$(C_{EQ})_{DEL}$	Delivered equipment cost	\$
C_{HE}	Purchased equipment cost, heat exchanger	\$
C_L	Cost of labor	\$
COE	Cash operating expenses	\$
C_P	Equipment cost in base year	\$
cP	Viscosity	cP
D	Depreciation	\$
DCFROR	Discounted cash flow rate of return	%
EBIT	Earnings before interest and taxes	\$
e	Naperian logarithm base	2.718
F	Future value, future worth, future amount	\$
F	Heat exchanger efficiency factor	Dimensionless
F_B	Heat exchanger design type	Dimensionless
FCI	Fixed capital investment	\$
FE	Fixed expenses	\$
FEL	Front-end loading	
FIFO	First in, first out (inventory)	lb
FOB	Free on board	
F_M	Material of construction cost factor	Dimensionless
F_P	Design pressure cost factor	Dimensionless
f_1, f_2, f_3	Inflation factors for years 1, 2, and 3	Dimensionless
f	Declining-balance factor	Dimensionless
I	Investment	\$
IRS	Internal Revenue Service	
i	Nominal interest	%
i_{eff}	Effective interest	%
K	Factor for cost index	Dimensionless
LIFO	Last in, last out (inventory)	lb
M	Annual raw material expense	\$
MACRS	Modified Accelerated Cost Recovery System	
m	Number of interest periods per year	Varies
m, n, p, q	constants or exponent	Dimensionless
N	Annual labor requirements	Operators per shift per year
n	Number of years, depreciation	Years
P	Principal, present value, present worth	\$
PC	Personal computer	
POP	Payout period (no interest)	Years
POP + I	Payout period plus interest	Years
Q	Energy transferred	Btu/h
R_1, R_2	Annual production rates	lb/yr
S	Salvage value or equipment capacity	Various
SL	Straight-line depreciation	
Sp gr	Specific gravity	Dimensionless
TE	Total expenses	\$
T_c	Combined incremental tax rate	%
T_f	Incremental federal income tax rate	%
T_s	Incremental state income tax rate	%
U	Annual utility expenses	\$
UAC	Uniform annual cost	\$
U_D	Overall heat-transfer coefficient	$Btu/(h \cdot ft^2 \cdot °F)$
V_e	Asset value at end of year	\$
V_i	Asset value at beginning of year	\$
VE	Variable expenses	\$
VIP	Value-improving practice	
X	Plant capacity	tons/day
Y	Operating labor	operator-hour/ton per processing step

General References: Allen, D. H., *Economic Evaluation of Projects*, 3d ed., Institution of Chemical Engineers, Rugby, England, 1991. Baasel, W. D., *Chemical Engineering Plant Design*, 2d ed., Van Nostrand Reinhold, New York, 1989. Brown, T. R., Hydrocarbon Processing, October 2000, pp. 93–100. Canada, J. R., and J. A. White, *Capital Investment Decision: Analysis for Management and Engineering*, 2d ed., Prentice-Hall, Englewood Cliffs, N.J., 1980. *Chemical Engineering* (ed.), *Modern Cost Engineering*, McGraw-Hill, New York, 1979. Couper, J. R., and W. H. Rader, *Applied Finance and Economic Analysis for Scientists and Engineers*, Van Nostrand Reinhold, New York, 1986. Couper, J. R., and O. T. Beasley, *The Chemical Process Industries: Function and Economics*, Dekker, New York, 2001. Couper, J. R., *Process Engineering Economics*, Dekker, New York, 2003. Garrett, D. E., *Chemical Engineering Economics*, Van Nostrand Reinhold, New York, 1989. Grant, E. L., and W. G. Ireson, *Engineering Economy*, 2d ed., Wiley, New York, 1950. Grant, E. L, W. G. Ireson, and R. S. Leavenworth, *Engineering Economy*, 8th ed., Wiley, New York, 1990. Hackney, J. W., and K. K. Humphreys (eds.), *Control and Management of Capital Projects*, 2d ed., McGraw-Hill, New York, 1992. Hill, D. A., and L. E. Rockley, *Secrets of Successful Financial Management*, Heinemann, London, 1990. Holland, F. A., F. A. Watson, and J. E. Wilkerson, *Introduction to Process Economics*, 2d ed., Wiley, London, 1983. K. K. Humphreys, F. C. Jelen, and J. H. Black (eds.), *Cost and Optimization Engineering*, 3d ed., McGraw-Hill, New York, 1991. *A Guide to Capital Cost Estimation*, 3d ed., Institution of Chemical Engineers, Rugby, England, 1990. Kharbanda, O. P., and E. A. Stallworthy, *Capital Cost Estimating in the Process Industries*, 2d ed., Butterworth-Heinemann, London, 1988. *How to Read an Annual Report*, Merrill Lynch, New York, 1997. Nickerson, C. B., *Accounting Handbook for Non Accountants*, 2d ed., CBI Publishing, Boston, 1979. Ostwald, P. F., *Engineering Cost Estimating*, 3d ed., Prentice-Hall, Englewood Cliffs, N.J., 1991. Park, W. R., and D. E. Jackson, *Cost Engineering Analysis*, 2d ed., Wiley, New York, 1984. Peters, M. S., and K. D. Timmerhaus, *Plant Design and Economics for Chemical Engineers*, 6th ed., McGraw-Hill, New York, 2003. Popper, H. (ed.), *Modern Cost Estimating Techniques*, McGraw-Hill, New York, 1970. Rose, L. M., *Engineering Investment Decisions: Planning under Uncertainty*, Elsevier, Amsterdam, 1976. Thorne, H. C., and J. B. Weaver (eds.), *Investment Appraisal for Chemical Engineers*, American Institute of Chemical Engineers, New York, 1991. Ulrich, G., and P. T. Vasudevan, *Chemical Engineering Process Design and Economics*, CRC Press, Boca Raton, Fla., 2004. Valle-Riestra, J. F., *Project Evaluation in the Chemical Process Industries*, McGraw-Hill, New York, 1983. Wells, G. L., *Process Engineering with Economic Objectives*, Wiley, New York, 1973. Woods, D. R., *Process Design and Engineering*, Prentice-Hall, Englewood Cliffs, N.J., 1993.

GENERAL COMMENTS

One of the most confusing aspects of process engineering economics is the nomenclature used by various authors and companies. In this part of Sec. 9, generic, descriptive terms have been used. Further, an attempt has been made to bring together most of the methods currently in use for project evaluation and to present them in such a way as to make them amenable to modern computational techniques. Most of the calculations can be performed on handheld calculators equipped with scientific function keys. For calculations requiring greater sophistication than that of handheld calculators, algorithms may be solved by using such programs as MATHCAD, TKSOLVER, etc. Spreadsheets are also used whenever the solution to a problem lends itself to this technique.

The nomenclature in process economics has been developed by accountants, engineers, and others such that there is no one correct set of nomenclature. Often it seems confusing, but one must question what is meant by a certain term since companies have adopted their own language. A glossary of terms is included at the end of this section to assist the reader in understanding the nomenclature. Further, abbreviations of terms such as DCFRR (discounted cash flow rate of return) are used to reduce the wordiness. The number of letters and numbers used to define a variable has been limited to five. The parentheses are removed whenever the letter group is used to define a variable for a computer. Also, a general symbol is defined for a type variable and is modified by mnemonic subscript, e.g., an annual cash quantity, annual capital outlay A_{TC}, \$/year. Wherever a term like this is introduced, it is defined in the text.

It is impossible to allow for all possible variations of equation requirements, but it is hoped that the nomenclature presented will prove adequate for most purposes and will be capable of logical extension to other more specialized requirements.

ACCOUNTING AND FINANCIAL CONSIDERATIONS

PRINCIPLES OF ACCOUNTING

Accounting has been defined as the art of recording business transactions in a systematic manner. It is the language of business and is used to communicate financial information. Conventions that govern accounting are fairly simple, but their application is complex. In this section, the basic principles are illustrated by a simple example and applied to analyzing a company report. The fair allocation of costs requires considerable technical knowledge of operations, so a close liaison between process engineers and accountants in a company is desirable.

In simplest terms, assets that are the economic resources of a company are balanced against equities that are claims against the firm. In equation form,

$$\text{Assets} = \text{Equities}$$

or

$$\text{Assets} = \text{Liabilities} + \text{Owners' Equity}$$

This dual aspect has led to the *double-entry* bookkeeping system in use today. Any transaction that takes place causes changes in the accounting equation. An increase in assets must be accompanied by one of the following:

- An increase in liabilities
- An increase in stockholders' equity
- An increase in assets

A change in one part of the equation due to an economic transaction must be accompanied by an equal change in another place—therefore, the term *double-entry* bookkeeping. On a page of an account, the left-hand side is designated the *debit* side and the right-hand side is the *credit* side. This convention holds regardless of the type of account. Therefore, for every economic transaction, there is an entry on the debit side balanced by the same entry on the credit side.

All transactions in their original form (receipts and invoices) are recorded chronologically in a *journal*. The date of the transaction together with an account title and a brief description of the transaction is entered. Table 9-1 is an example of a typical journal page for a company. Journal entries are transferred to a *ledger* in a process called *posting*. Separate ledger accounts, such as a revenue account, expense account, liability account, or asset account, may be set up for each major transaction. Table 9-2 shows an example of a typical ledger page. The number of ledger accounts depends on the information that management needs to make decisions. Periodically, perhaps on a monthly basis but certainly on a yearly basis, the ledger sheets are closed and balanced. The ledger sheets are then intermediate documents between journal records and balance sheets, income statements, and retained earnings statements, and they provide information for management and various government reports. For example, a consolidated income statement can be prepared for the ledger, revenue, and expense accounts. In like manner, the asset and liability accounts provide information for balance sheets.

TABLE 9-1 Typical Journal Page

Date	Explanation	LP	Debit	Credit
200X				
Mar 1	Cash	1	$95,000	
	J. Jones, Capital	2		$95,000
Mar 4	Property	4	5,000	
	Cash	1		3,000
	Mortgage	3		2,000
Mar 11	Remodeling Bldg.	5	7,800	
	Cash	1		7,800
Mar 13	Equipment	6	62,300	
	Cash	1		10,000
	Note Payable	3		52,300
Apr 4	To J. Jones	2	2,500	
	Cash	1		2,500

SOURCE: J. R. Couper, *Process Engineering Economics*, Dekker, New York, 2003. By permission of Taylor & Francis Books, Inc., Boca Raton, Fla.

FINANCIAL STATEMENTS

A basic knowledge of accounting and financial statements is necessary for a chemical professional to be able to analyze a firm's operation and to communicate with accountants, financial personnel, and managers. Financial reports of a company are important sources of information used by management, owners, creditors, investment bankers, and financial analysts. All publicly held companies are required to submit annual reports to the Securities and Exchange Commission. As with any field a certain basic nomenclature is used to be able to understand the financial operation of a company. It should be emphasized that companies may also have their own internal nomenclature, but some terms are universally accepted. In this section, the common terminology is used.

A financial report contains two important documents—the *balance sheet* and the *income statement*. Two other documents that appear in the financial report are the *accumulated retained earnings* and the *changes in working capital*. All these documents are discussed in the following sections using a fictitious company.

Balance Sheet The balance sheet represents an accounting view of the financial status of a company on a particular date. Table 9-3 is an example of a balance sheet for a company. The date frequently used by corporations is December 31 of any given year, although some companies are now using June 30 or September 30 as the closing date. It is as if the company's operation were frozen in time on that date. The term *consolidated* means that all the balance sheet and income statement data include information from the parent as well as subsidiary operations. The balance sheet consists of two parts: *assets* are the items that the company owns, and *liabilities and stockholders' equity* are what the company owes to creditors and stockholders. Although the balance sheet has two sides, it is not part of the double-entry accounting system. The balance sheet is not an account but a statement of claims against company assets on the date of the reporting period. The claims are the creditors and the stockholders. Therefore, the total assets must equal the total liabilities plus the stockholders' equity.

Assets are classified as current, fixed, or intangibles. *Current assets* include cash, cash equivalents, marketable securities, accounts receivable, inventories, and prepaid expenses. Cash and cash equivalents are those items that can be easily converted to cash. Marketable securities are securities that a company holds that also may be converted to cash. Accounts receivable are the amounts due a company from customers from material that has been delivered but has not been collected as yet. Customers are given 30, 60, or 90 days in which to pay; however, some customers fail to pay bills on time or may not be able to pay at all. An allowance is made for doubtful accounts. The amount is deducted from the accounts receivables. Inventories include the cost of raw materials, goods in process, and product on hand. Prepaid expenses include insurance premiums paid, charges for leased equipment, and charges for advertising that are paid prior to the receipt of the benefit from these items. The sum of all the above items is the *total current assets*. The term *current* refers to the fact that these assets are easily converted within a year, or more likely in a shorter time, say, 90 days.

Fixed assets are items that have a relatively long life such as land, buildings, and manufacturing equipment. The sum of these items is the *total property, plant, and equipment*. From this total, *accumulated depreciation* is subtracted and the result is *net property and equipment*. Last, an item referred to as *intangibles* includes a variety of items such as patents, licenses, intellectual capital, and goodwill. Intangibles are difficult to evaluate since they have no physical existence; e.g., goodwill is the value of the company's name and reputation. The sum of the *total current assets, net property*, and *intangibles* is the *total assets*.

Liabilities are the obligations that the company owes to creditors and stockholders. *Current liabilities* are obligations that come due within a year and include *accounts payable* (money owed to creditors for goods and services), *notes payable* (money owed to banks, corporations, or other lenders), *accrued expenses* (salaries and wages to employees, interest on borrowed funds, fees due to professionals, etc.), *income taxes payable, current part of long-term debt*, and *other current liabilities* due within the year.

Long-term liabilities are the amounts due after 1 year from date of the financial report. They include *deferred income taxes* that a company is permitted to postpone due to accelerated depreciation to encourage investment, (but they must be paid sometime in the future) and *bonds and notes* that do not have to be paid within the year but at some later date. The sum of the *current* and *long-term liabilities* is the *total liabilities*.

TABLE 9-2 Typical Ledger Page

			Cash: Account 01			
200X Mar 1 Capital	J-1	$95,000	Mar 1 Mar 11 Mar 13 Apr 4	Property Remodeling Equipment J. Jones	J-1 J-1 J-1 J-1	$3,000 7,800 10,000 2,500
			Capital: Account 02			
Apr 4 Cash to J. Jones	J-1	$2,500	Mar 1	Capital	J-1	$95,000
			Accounts Payable: Account 03			
			Mar 4 Mar 13	Mortgage Note Payable	J-1 J-1	$2,000 52,300
			Property and Building: Account 04			
Mar 4 Mar 11	J-1 J-1	$5,000 7,800				
			Equipment: Account 05			
Mar 13	J-1	$62,300				

SOURCE: J. R. Couper, *Process Engineering Economics*, Dekker, New York, 2003. By permission of Taylor & Francis Books, Inc., Boca Raton, Fla.

TABLE 9-3 Consolidated Balance Sheet[a] (December 31)

Assets	2005	2004
Current assets		
Cash	$63,000	$51,000
Marketable securities	41,000	39,000
Accounts receivable[b]	135,000	126,000
Inventories	149,000	153,000
Prepaid expenses	3,200	2,500
Total current assets	$391,200	$371,500
Fixed assets		
Land	35,000	35,000
Buildings	101,000	97,500
Machinery	278,000	221,000
Office equipment	24,000	19,000
Total fixed assets	$438,000	$372,500
Less accumulated depreciation	128,000	102,000
Net fixed assets	$310,000	$270,500
Intangibles	4,500	4,500
Total assets	$705,700	$646,500
Liabilities	2005	2004
Current liabilities		
Accounts payable	$92,300	$81,300
Notes payable	67,500	59,500
Accrued expenses payable	23,200	26,300
Federal income taxes payable	18,500	17,500
Total current liabilities	$201,500	$184,600
Long-term liabilities		
Debenture bonds, 10.3% due in 2015	110,000	110,000
Debenture bonds, 11.5% due in 2007	125,000	125,000
Deferred income taxes	11,600	10,000
Total liabilities	$448,100	$429,600
Stockholder's equity		
Preferred stock, 5% cumulative		
$5 par value—200,000 shares	$10,000	$10,000
Common stock, $1 par value		
2000 28,000,000 shares	32,000	28,000
2000X 32,000,000 shares		
Capital surplus	8,000	6,000
Accumulated retained earnings	207,600	172,900
Total stockholder's equity	**$257,600**	**$216,900**
Total liabilities and stockholder's equity	**$705,700**	**$646,500**

[a]All amounts in thousands of dollars.
[b]Includes an allowance for doubtful accounts.
SOURCE: J. R. Couper, *Process Engineering Economics*, Dekker, New York, 2003. By permission of Taylor & Francis Books, Inc., Boca Raton, Fla.

Stockholders' equity is the interest that all stockholders have in a company and is a liability with respect to the company. This category includes *preferred* and *common stock* as well as additional paid-in capital (the amount that stockholders paid above the par value of the stock) and *retained earnings.* These are earnings from accumulated profit that a company earns and are used for reinvestment in the company. The sum of these items is the *stockholders' equity.*

On a balance sheet, the sum of the *total liabilities* and the *stockholders' equity* must equal the *total assets,* hence the term *balance sheet.* Comparing balance sheets for successive years, one can follow changes in various items that will indicate how well the company manages its assets and meets its obligations.

Income Statement An income statement shows the revenue and the corresponding expenses for the year and serves as a guide for how the company may do in the future. Often income statements may show how the company performed for the last two or three years. Table 9-4 is an example of a consolidated income statement.

Net sales are the primary source of revenue from goods and services. This figure includes the amount reported after returned goods, discounts, and allowances for price reductions are taken into account. *Cost of sales* represents all the expenses to convert raw materials to finished products. The major components of these expenses are direct material, direct labor, and overhead. If the cost of sales is subtracted from net sales, the result is the *gross margin.* One of the most important items on the income statement is *depreciation* and *amortization.* Depreciation is an allowance the federal government permits for the

TABLE 9-4 Consolidated Income Statement (December 31)

	2005	2004
Net sales (revenue)	$932,000	$850,000
Cost of sales and operating expenses		
Cost of goods sold	692,000	610,000
Depreciation and amortization	40,000	36,000
Sales, general, and administrative expenses	113,500	110,000
Operating profit	$86,500	$94,000
Other income (expenses)		
Dividends and interest income	10,000	7,000
Interest expense	(22,000)	(22,000)
Income before provision for income taxes	$74,500	$79,000
Provision for federal income taxes	24,500	26,000
Net profit for year	$50,000	$53,000

SOURCE: J. R. Couper, *Process Engineering Economics*, Dekker, New York, 2003. By permission of Taylor & Francis Books, Inc., Boca Raton, Fla.

wear and tear as well as the obsolescence of plant and equipment and is treated as an expense. Amortization is the decline in value of intangible assets such as patents, franchises, and goodwill. *Selling, general, and administrative expenses* include the marketing salaries, advertising expenses, travel, executive salaries, as well as office and payroll expenses. When depreciation, amortization, and the sales and administrative expenses are subtracted from the gross margin, the result is the *operating income. Dividends and interest income* received by the company are then added. Next *interest expense* earned by the stockholders and *income taxes* are subtracted, yielding the term *income before extraordinary loss.* It is the expenses a company may incur for unusual and infrequent occasions. When all the above items are added or subtracted from the operating income, *net income* (or *loss*) is obtained. This latter term is the "bottom line" often referred to in various reports.

Accumulated Retained Earnings This is an important part of the financial report because it shows how much money has been retained for growth and how much has been paid as dividends to stockholders. When the accumulated retained earnings increase, the company has greater value. The calculation of this value of the retained earnings begins with the previous year's balance. To that figure add the net profit after taxes for the year. Dividends paid to stockholders are then deducted, and the result is the accumulated retained earnings for the year. See Table 9-5.

Concluding Remarks One of the most important sections of an annual report is the "notes." These contain any liabilities that a company may have due to impending litigation that could result in charges or expenses not included in the annual report.

OTHER FINANCIAL TERMS

Profit margin is the ratio of net income to total sales, expressed as a percentage or sometimes quoted as the ratio of profit before interest and taxes to sales, expressed as a percentage. *Operating margin* is obtained by subtracting operating expenses from gross profit expressed as a percentage of sales. *Net worth* is the difference between total assets and total liabilities plus stockholders' equity. *Working capital* is the difference between total current assets and current liabilities.

TABLE 9-5 Accumulated Retained Earnings Statement[a] (December 31)

	2005	2004
Balance as of January 1	$172,900	$141,850
Net profit for year	50,000	53,000
Total for year	$222,900	$194,850
Less dividends paid on:		
Preferred stock	700	700
Common stock	14,600	21,250
Balance December 31	$207,600	$172,900

[a]All amounts in thousands of dollars.
SOURCE: J. R. Couper, *Process Engineering Economics*, Dekker, New York, 2003. By permission of Taylor & Francis Books, Inc., Boca Raton, Fla.

FINANCIAL RATIOS

There are many financial ratios of interest to financial analysts. A brief discussion of some of these ratios follows; however, a more complete discussion may be found in Couper (2003).

Liquidity ratios are a measure of a company's ability to pay its short-term debts. *Current ratio* is obtained by dividing the current assets by the current liabilities. Depending on the economic climate, this ratio is 1.5 to 2.0 for the chemical process industries, but some companies operate closer to 1.0. The *quick ratio* is another measure of liquidity and is cash plus marketable securities divided by the current liabilities and is slightly greater than 1.0.

Leverage ratios are an indication of the company's overall debt burden. The *debt/total assets ratio* is determined by dividing the total debt by total assets expressed as a percentage. The industry average is 35 percent. *Debt/equity ratio* is another such ratio. The higher these ratios, the greater the financial risk since if an economic downturn did occur, it might be difficult for a company to meet the creditors' demands. The *times interest earned* is a measure of the extent to which profit could decline before a company is unable to pay interest charges. The ratio is calculated by dividing the *earnings before interest and taxes (EBIT)* by interest charges. The *fixed-charge coverage* is obtained by dividing the income available for meeting fixed charges by the fixed charges.

Activity ratios are a measure of how effectively a firm manages its assets. There are two inventory/turnover ratios in common use today. The *inventory/sales* ratio is found by dividing the inventory by the sales. Another method is to divide the cost of sales by inventory. The *average collection period* measures the number of days that customers' invoices remain unpaid. *Fixed assets* and *total assets turnover* indicate how well the fixed and total assets of the firm are being used.

Profitability ratios are used to determine how well income is being managed. The *gross profit margin* is found by dividing the gross profits by the net sales, expressed as a percentage. The *net operating margin* is equal to the earnings before interest and taxes divided by net sales. Another measure, the *profit margin on sales*, is calculated by dividing the net profit after taxes by net sales. The *return on total assets* ratio is the net profit after taxes divided by the total assets expressed as a percentage. The *return on equity* ratio is the net income after taxes and interest divided by stockholders' equity.

Table 9-6 shows the financial ratios for Tables 9-3 and 9-4. Table 9-7 is a summary of selected financial ratios and industry averages.

TABLE 9-6 Financial Ratios for Tables 9-3 and 9-4

Liquidity
Current ratio = \$391,200/\$201,500 = 1.94
Cash ratio = \$391,200 − 149,000/\$201,500 = 1.20
Leverage
Debt/assets ratio = [(\$448,100 − 201,500)/\$705,700] × 100 = 35%
Times interest earned = \$74,500 − 22,000/\$22,000 = 4.39
Fixed-charge coverage = \$86,500/\$22,000 = 3.93
Activity
Inventory turnover = \$932,000/\$149,000 = 6.25
Average collection period = \$135,000/(\$932,000/365) = 52.8 days
Fixed-assets turnover = \$932,000/\$438,000 = 2.13
Total-assets turnover = \$932,000/\$705,700 = 1.32
Profitability
Gross profit margin = [(\$932,000 − 692,000)/\$932,000] × 100 = 25.8%
Net operating margin = \$74,500/\$932,000 × 100 = 7.99%
Profit margin on sales = \$50,000/\$932,000 × 100 = 5.36%
Return on net worth (return on equity) = [\$50,000/(\$705,700 − 448,100)] × 100 = 19.4%
Return on total assets = (\$50,000/\$705,700) × 100 = 7.09%

RELATIONSHIP BETWEEN BALANCE SHEETS AND INCOME STATEMENTS

There is a relationship between these two documents because information obtained from each is used to calculate the returns on assets and equity. Figure 9-1 is an operating profitability tree for a fictitious company and contains the fixed and variable expenses as reported on internal company reports, such as the manufacturing expense sheet. Figure 9-2 is a financial family tree for the same company depicting the relationship between values in the income statement and the balance sheet.

FINANCING ASSETS BY DEBT AND/OR EQUITY

The various options for obtaining funds to finance new projects are not a simple matter. Significant factors such as the state of the economy, inflation, a company's present indebtedness, and the cost of capital will affect the decision. Should a company incur more long-term debt, or should it seek new venture capital from equity sources? A simple yes or no answer will not suffice because the financial decision is complex. One consideration is the company's position with respect to leverage. If a company has a large proportion of its debt in bonds and preferred stock, the common stock is highly leveraged. Should the earnings decline, say, by 10 percent, the dividends available to common stockholders might be wiped out. The company also might not be able to cover the interest on its bonds without dipping into the accumulated earnings. A high debt/equity ratio illustrates the fundamental weakness of companies with a large amount of debt. When low-interest financing is available, such as for large government projects, the return-on-equity evaluations are used. Such leveraging is tantamount to transferring money from one pocket to another; or, to

TABLE 9-7 Selected Financial Ratios

Item	Equation for calculation	Industry average
Liquidity		
Current ratio	Current assets/current liabilities	1.5–2.0
Cash ratio	Current assets − inventory/current liabilities	1.0–1.5
Leverage		
Debt to total assets	Total debt/total assets	30–40%
Times interest earned	Profit before taxes plus interest charges/interest charges	7.0–8.0
Fixed-charge coverage	Income available for meeting fixed charges/fixed charges	6.0
Activity		
Inventory turnover	Sales or revenue/inventory	7.0
Average collection period	Receivables/sales per day	40–60 days
Fixed assets turnover	Sales/fixed assets	2–4
Total assets turnover	Sales/total assets	1–2
Profitability		
Gross profit margin	Net sales − cost of goods sold/sales	25–40%
Net operating margin	Net operating profit before taxes/sales	10–15%
Profit margin on sales	Net profit after taxes/sales	5–8%
Return on net worth (return on equity)	Net profit after taxes/net worth	15%
Return on total assets	Net profit after taxes/total assets	7–10%

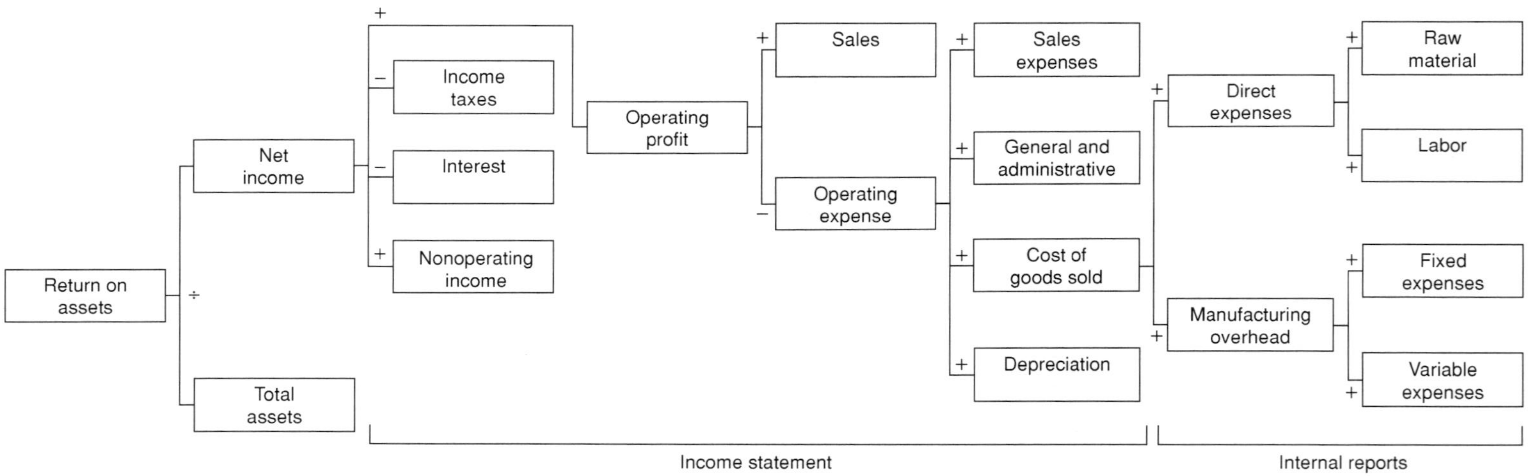

FIG. 9-1 Operating profitability tree. (*Source: Adapted from Couper, 2003.*)

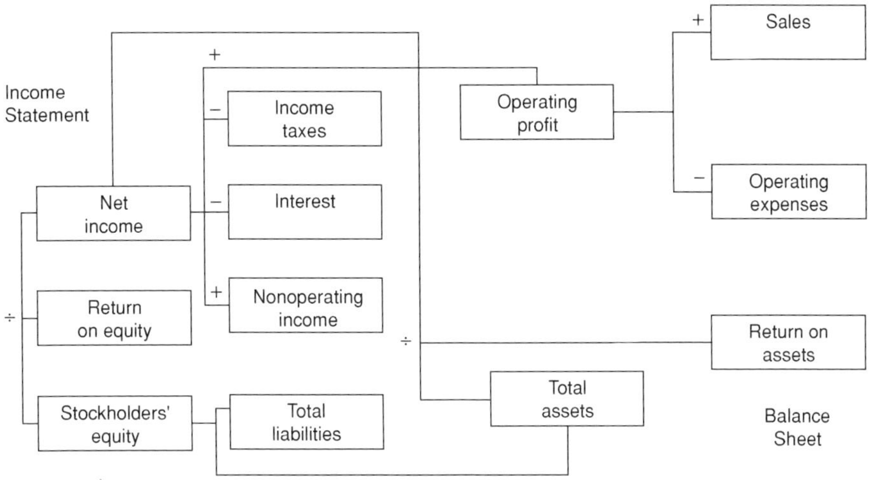

FIG. 9-2 Financial family tree. (*Source: Adapted from Couper, 2003.*)

put it another way, a company may find itself borrowing from itself. In the chemical process industries, debt/equity ratios of 0.3 to 0.5 are common for industries that are capital-intensive (Couper et al., 2001). Much has been written on the strategies of financing a corporate venture. The correct strategy has to be evaluated from the standpoint of what is best for the company. It must maintain a debt/equity ratio similar to those of successful companies in the same line of business.

COST OF CAPITAL

The cost of capital is what it costs a company to borrow money from all sources, such as loans, bonds, and preferred and common stock. It is an important consideration in determining a company's minimum acceptable rate of return on an investment. A company must make more than the cost of capital to pay its debts and make a profit. From profits, a company pays dividends to the stockholders. If a company ignores the cost of capital to increase dividends to the stockholders, then management is not meeting its obligations to pay off outstanding debts.

A sample calculation of the after-tax weighted cost of capital is found in Table 9-8. Each debt item is divided by the total debt, and that result is multiplied by the after-tax yield to maturity that equals the after-tax weighted average cost of that debt item contributing to the cost of capital. The information to estimate the cost of capital may be obtained from the annual report, the 10K, or the 10Q reports.

TABLE 9-8 Cost of Capital Illustration

Balance sheet 12/31/XX	Debt, $M	After-tax yield to maturity, %	After-tax weighted average cost, %
Long-term debt			
Revolving account	5.0	4.5	0.02
$4\frac{3}{8}$% debentures	12.0	4.0	0.05
$6\frac{1}{2}$% debentures	3.4	4.7	0.02
$6\frac{3}{4}$% debentures	9.4	4.2	0.04
$7\frac{1}{2}$% debentures	74.5	4.2	0.30
$9\frac{3}{8}$% loan	125.0	4.4	0.53
Other	23.2	4.4	0.10
Total long-term debt	252.5		1.06
Deferred taxes	67.7	0.0	0
Reserves	16.1	0.0	0
Preferred stock	50.0	8.6	0.42
Shareholders' equity	653.9	15.6	9.80
Total debt	1,040.2		11.28

Each debt item in $M divided by the total debt times the after-tax yield to maturity equals the after-tax weighted average cost contributing to the cost of capital.

SOURCE: Private communication.

WORKING CAPITAL

The accounting definition of *working capital* is total current assets minus total current liabilities. This information can be found from the balance sheet. Current assets consist chiefly of cash, marketable securities, accounts receivable, and inventories; current liabilities include accounts payable, short-term debts, and the part of the long-term debt currently due. The accounting definition is in terms of the entire company.

For economic evaluation purposes, another definition of working capital is used. It is the funds, in addition to the fixed capital, that a company must contribute to a project. It must be adequate to get the plant in operation and to meet subsequent obligations when they come due. Working capital is not a one-time investment that is known at the project inception, but varies with the sales level and other factors. The relationship of working capital to other project elements may be viewed in the cash flow model (see Fig. 9-9). Estimation of an adequate amount of working capital is found in the section "Capital Investment."

INVENTORY EVALUATION AND COST CONTROL

Under ordinary circumstances, inventories are priced (valued) at some form of cost. The problem in valuating inventory lies in "determining what costs are to be identified with inventories in a given situation" (Nickerson, 1979).

Valuation of materials can be made by using the

- Cost of a specific lot
- Average cost
- Standard cost

Under "cost of a specific lot," those lots to be valuated must be identified by referring to related invoices. Many companies use the average cost for valuating inventories. The average used should be weighted by the quantities purchased rather than by an average purchase price. Average cost method tends to spread the effects of short run price changes and has a tendency to level out profits in those industries that use raw materials whose prices are volatile. For many manufacturing companies, inventory valuation is an important consideration varying in degree of importance. Inventories that are large are subject to significant fluctuations from time to time in size and mix and in prices, costs, and values.

Materials are valuated in accordance with their acquisition. Some companies use the first-in, first-out (FIFO) basis. Materials are used in order of their acquisition to minimize losses from deterioration. Another method is last-in, first-out (LIFO) in which materials coming in are the first to leave storage for use. The method used depends on a number of factors. Accounting texts discuss the pros and cons of each method, often giving numerical examples. Some items to consider are income tax considerations and cash flow that motivate management to adopt conservative valuation policies. Tax savings may accrue using one method compared to the other, but they may not be permanent. Whatever method is selected, consistency is important so that comparability of reported figures may be maintained from one time period to another. It is management's responsibility to make the decision regarding the method used. In some countries, government regulations control the method to be used. There are several computer software programs that permit the user to organize, store, search, and manage inventory from a desktop computer.

BUDGETS AND COST CONTROL

A budget is an objective expressed in monetary terms for planning and controlling the resources of a company. Budgeted numbers are objectives, not achievements. A comparison of actual expenses with budgeted (cost standards) figures is used for control at the company, plant, departmental, or project level. A continuing record of performance should be maintained to provide the data for preparing future budgets (Nickerson, 1979). Often when a company compares actual results with cost standards or budgeted figures, a need for improving operations will surface. For example, if repairs to equipment continuously exceed the budgeted amount, perhaps it is time to consider replacement of that older equipment with a newer, more efficient model. Budgets are usually developed for a 1-year period; however, budgets for various time frames are frequently prepared. For example, in planning future operations, an intermediate time period of, say, 5 years may be appropriate, or for long-range planning the time period selected may be 10 years.

A cost control system is used

- To provide early warning of uneconomical or excessive costs in operations
- To provide relevant feedback to the personnel responsible for developing budgets
- To develop cost standards
- To promote a sense of cost consciousness
- To summarize progress

Budgetary models based upon mathematical equations are available to determine the effect of changes in variables. There are numerous sources extant in the literature for these models.

CAPITAL COST ESTIMATION

TOTAL CAPITAL INVESTMENT

The *total capital investment* includes funds required to purchase land, design and purchase equipment, structures, and buildings as well as to bring the facility into operation (Couper, 2003). The following is a list of items constituting the total capital investment:

Land
Fixed capital investment
Offsite capital
Allocated capital
Working capital
Start-up expenses
Other capital items (interest on borrowed funds prior to start-up; catalysts and chemicals; patents, licenses, and royalties; etc.)

Land Land is often acquired by a company some time prior to the building of a manufacturing facility. When a project is committed to be built on this land, the value of the land becomes part of that facility's capital investment.

Fixed Capital Investment When a firm considers the manufacture of a product, a capital cost estimate is prepared. These estimates are required for a variety of reasons such as feasibility studies, the selection of alternative processes or equipment, etc., to provide information for planning capital appropriations or to enable a contractor to bid on a project. Included in the fixed capital investment is the cost of purchasing, delivery, and installation of manufacturing equipment, piping, automatic controls, buildings, structures, insulation, painting, site preparation, environmental control equipment, and engineering and construction costs. The fixed capital investment is significant in developing the economics of a process since this figure is used in estimating operating expenses and calculating depreciation, cash flow, and project profitability. The estimating method used should be the best, most accurate means consistent with the time and money available to prepare the estimate.

Classification of Estimates There are two broad classes of estimates: grass roots and battery limits. Grass-roots estimates include the entire facility, starting with site preparation, buildings and structures, processing equipment, utilities, services, storage facilities, railroad yards, docks, and plant roads. A battery-limits estimate is one in which an imaginary boundary is drawn around the proposed facility to be estimated. It is assumed that all materials, utilities, and services are available in the quality and quantity required to manufacture a product. Only costs within the boundary are estimated.

Quality of Estimates Capital cost estimation is more art than science. An estimator must use considerable judgment in preparing the estimate, and as the estimator gains experience, the accuracy of the estimate improves. There are several types of fixed capital cost estimates:

- *Order-of-magnitude (ratio estimate).* Rule-of-thumb methods based on cost data from similar-type plants are used. The probable accuracy is –30 percent to +50 percent.
- *Study estimate (factored estimate).* This type requires knowledge of preliminary material and energy balances as well as major equipment items. It has a probable accuracy of –25 to +30 percent.
- *Preliminary estimate (budget authorization estimate).* More details about the process and equipment, e.g., design of major plant items, are required. The accuracy is probably –20 to +25 percent.
- *Definitive estimate (project control estimate).* The data needed for this type of estimate are more detailed than those for a preliminary estimate and include the preparation of specifications and drawings. The probable accuracy is –10 to +15 percent.
- *Detailed estimate (firm estimate).* Complete specifications, drawings, and site surveys for the plant construction are required, and the estimate has an accuracy of –5 to +10 percent.

Detailed information requirements for each type of estimate may be found in Fig. 9-3.

In periods of high inflation, the results of various estimates and accuracy may overlap. At such times, four categories may be more suitable, namely, study, preliminary, definitive, and detailed categories. At present, some companies employing the *front-end loading (FEL)* process for project definition and execution use three categories:

Project stage	Accuracy
Conceptual	±/40%
Feasibility	±/25%
Definition	±/10%

For more information on the FEL process, see "Capital Project Execution and Analysis" near the end of Sec 9.

Scope The scope is a document that defines a project. It contains words, drawings, and costs. A scope should answer the following questions clearly:

What product is being manufactured?
How much is being produced?

ESTIMATING INFORMATION GUIDE

Information Either Required or Available

		Order of magnitude (ratio)	Study (factored)	Preliminary (budget authorization)	Definitive (project control)	Detailed (firm)
Site	Location		•	•	•	•
	General description		•	•	•	•
	Site survey			•	•	•
	Geotechnical report			•	•	•
	Site plot plan and contours				•	•
	Well-developed site facilities					•
Process flow	Rough sketches		•			
	Preliminary			•		
	Engineered				•	•
Equipment	Rough sizes and construction		•	•		
	Engineered specifications				•	•
	Vessel data sheets				•	•
	General arrangement			•	•	
	Final arrangement					•
Buildings and structures	Rough sizes and construction		•	•		
	Foundation sketches			•	•	
	Architectural and construction			•	•	•
	Preliminary structural design			•	•	
	General arrangements and elevations				•	•
	Detailed drawings					•
Utilities and services	Rough quantities		•			
	Preliminary heat balance			•		
	Preliminary flow sheets			•		
	Engineered heat balance				•	•
	Engineered flow sheets				•	•
	Detailed drawings					•
Piping and insulation	Preliminary flow sheets		•	•		
	Engineered flow sheets				•	•
	Piping layouts and schedules					•
	Insulation rough specifications			•		
	Insulation applications				•	•
	Insulation details					•
Instrumentation	Preliminary list			•		
	Engineered list				•	•
	Detail drawings					•
Electrical	Rough motor list and sizes		•	•		
	Engineered list and sizes				•	•
	Substation number and size			•	•	•
	Preliminary specifications			•	•	
	Distribution specifications					•
	Preliminary interlocks and controls				•	
	Engineered single-line diagrams				•	•
	Detailed drawings					•
Work-hours	Engineering and drafting		•	•	•	•
	Construction supervision					•
	Craft labor				•	•
Project scope	Product, capacity, location, utilities, and services; Building requirements, process, storage, and handling	•	•	•	•	•

FIG. 9-3 Information guide for preparing estimates. (*Source*: Perry's Chemical Engineers' Handbook, *5th ed., McGraw-Hill, New York, 1973.*)

What is the quality of the product?
Where is the product to be produced?
What is the quality of the estimate?
What is the basis for the estimate?
What are the knowns and unknowns with respect to the project?

Before an estimate can be prepared, it is essential to prepare a scope. It may be as simple as a single page, such as for an order-of-magnitude estimate, or several large manuals, for a detailed estimate. As the project moves forward from inception to a detailed estimate, the scope must be revised and updated to provide the latest information. Changes during the progress of a project are inevitable, but a well-defined scope prepared in advance can help minimize costly changes. If a scope is properly defined, the following results:

An understanding between those who prepared the scope (engineering) and those who accept it (management)
A document that indicates clearly what is provided in terms of technology, quality, schedule, and cost
A basis in enough detail to be used in controlling the project and its costs to permit proper evaluation of any proposed changes
A device to permit subsequent evaluation of the performance compared to the intended performance
A document to control the detailed estimate for the final design, construction, and design

Equipment Cost Data The foundation of a fixed capital investment estimate is the equipment cost data. From this information, through the application of factors or percentages based upon the estimator's experience, the fixed capital investment is developed.

These cost data are reported as purchased, delivered, or installed cost. Purchased cost is the price of the equipment FOB at the manufacturer's plant, whereas delivered cost is the purchased price plus the delivery charge to the purchaser's plant FOB. Installed cost means the equipment has been purchased, delivered, uncrated, and placed on a foundation in the purchaser's operating department but does not include piping, electrical, instrumentation, insulation, etc., costs. Perhaps a better name might be *set-in-place* cost.

It is essential to have reliable cost data since the engineer producing the estimate starts with this information and develops the fixed capital cost estimate. The estimator must know the source of the data, the basis for the data, its date, potential errors, and the range over which the data apply. There are many sources of graphical equipment cost data in the literature, but some are old and the latest published data were in the early 1990s. There have been no significant cost data published recently. To obtain current cost data, one should solicit bids from vendors; however, it is essential to impress on the vendor that the information is to be used for preliminary estimates. A disadvantage of using vendor sources is that there is a chance of compromising proprietary information.

Cost-capacity plots of equipment indicate a straight-line relationship on a log-log plot. Figure 9-4 is an example of such a plot. A convenient method of presenting these data is in equation format:

$$C_2 = C_1\left(\frac{S_2}{S_1}\right)^n \qquad (9\text{-}1)$$

where C_1 = cost of equipment of capacity S_1
C_2 = cost of equipment of capacity S_2

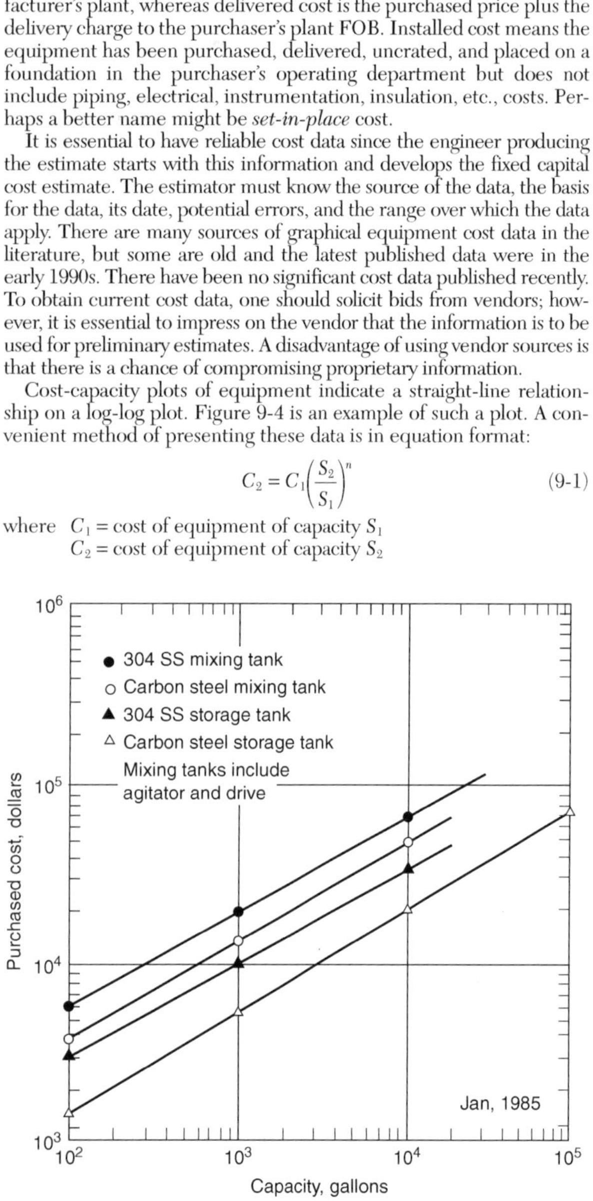

FIG. 9-4 Cost-capacity plot.

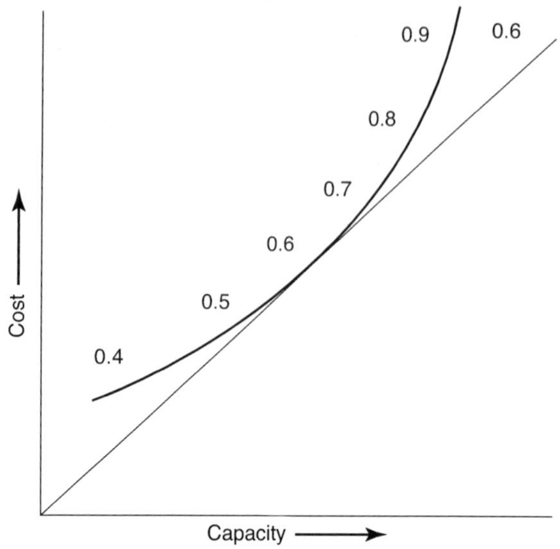

FIG. 9-5 Variation of n on cost-capacity plot.

n = exponent that may vary between 0.4 and 1.2 depending on type of equipment

Equation (9-1) is known as the *six-tenths rule* since the average value for all equipment is about 0.6. D. S. Remer and L. H. Chai (*Chemical Engineering Progress*, August 1990, pp. 77–82) published an extensive list of six-tenths data. Figure 9-5 shows how the exponent may vary from 0.4 to 0.9 for a given equipment item. Data accuracy is the highest in the narrow, middle-range of capacity, but at either end of the plot, the error is great. These errors occur when one correlates cost data with one independent variable when more than one variable is necessary to represent the data, or when pressure, temperature, materials of construction, or design features vary considerably.

A convenient way to display cost-capacity data is by algorithms. They are readily adaptable for computerized cost estimation programs. Algorithm modifiers in equation format may be used to account for temperature, pressure, material of construction, equipment type, etc. Equation (9-2) is an example of obtaining the cost of a shell-and-tube heat exchanger by using such modifiers.

$$C_{HE} = KC_BF_DF_MF_P \qquad (9\text{-}2)$$

where C_{HE} = purchased equipment cost
K = factor for cost index based upon a base year
C_B = base cost of a carbon-steel floating-head exchanger, 150-psig design pressure
F_D = design-type cost factor if different from that in C_B
F_M = material-of-construction cost factor
F_P = design pressure cost factor

Each cost factor is obtained from equations or tables from Couper, 2003, App. C, and have been updated to third-quarter 2002.

Cost Indices Cost data are given as of a specific date and can be converted to more recent costs through the use of cost indices. In general, the indices are based upon constant dollars in a base year and actual dollars in a specific year. In this way, with the proper application of the index, the effect of inflation (or deflation) and price increases by multiplying the historical cost by the ratio of the present cost index divided by the index applicable in the historical year. Labor, material, construction costs, energy prices, and product prices all change at different rates. Most cost indices represent national averages, and local averages may vary considerably. Table 9-9 is a list of selected values of three cost indices of significance in the chemical process industries.

TABLE 9-9 Selected Cost Indices

Year Base	M&S Index[(1)a] 1926 = 100	CE index[(1)] 1957 – 1959 = 100	Nelson-Farrar index[(2)] 1946 = 100
1990	915.1	357.6	1225.7
1992	943.1	358.2	1277.3
1994	964.2	368.1	1349.7
1996	1039.1	381.7	1418.9
1998	1061.9	389.5	1477.6
2000	1089.0	394.1	1542.7
2001	1093.9	394.3	1579.7
2002	1104.2	395.6	1642.2
2003	1123.6	402.0	1710.4
2004 (3Q)	1124.7	457.4	1856.1

(1) From 1990 onward, the M&S and CE indices are from *Chemical Engineering* magazine.
(2) The Nelson-Farrar indices from 1990 onward are found in *Oil and Gas Journal*.
[a]Process industry average instead of all industry average.

The chemical engineering (CE) index and the Marshall and Swift index are found in each issue of the magazine *Chemical Engineering*. The *Oil and Gas Journal* reports the *Nelson-Farrar Refinery* indices in the first issue of each quarter. The base years selected for each index are generally periods of low inflation so that the index is stable. The derivation of base values is referred to in the respective publications.

A cost index is used to project a cost from a base year to another selected year. The following equation is used:

$$\text{Cost at } \Theta_2 = \text{cost at } \Theta_2 \left(\frac{\text{index at } \Theta_2}{\text{index at } \Theta_2}\right) \tag{9-3}$$

Example 1: Use of Cost Index A centrifuge cost \$95,000 in 1999. What is the cost of the same centrifuge in third quarter of 2004? Use the CE index.

Solution:

CE index in 1999 = 390.6

CE index in 3d quarter 2004 = 457.4

Cost in 2004 = cost in 1999 (CE index in 3d quarter 2004/ CE index in 1999)

$$= \$95{,}000\left(\frac{457.4}{390.6}\right) = \$111{,}200$$

Inflation When costs are to be projected into the future due to inflation, it is a highly speculative exercise, but it is necessary for estimating investment costs, operating expenses, etc. Inflation is the increase in price of goods without a corresponding increase in productivity. A method for estimating an inflated cost is

$$C_i = (1 + f_1)(1 + f_2)(1 + f_3)\, C_P \tag{9-4}$$

where C_i = inflated cost
f_1 = inflation rate the first year
f_2 = inflation rate the second year
f_3 = inflation rate the third year
C_P = cost in a base year

The assumed inflation factors f are obtained from federal economic reports, financial sources such as banks and investment houses, and news media. These factors must be reviewed periodically to update estimates.

Example 2: Inflation A dryer today costs \$475,000. The projected inflation rates for the next 3 years are 3, 4.2, and 4.7 percent. Calculate the projected cost in 3 years.

Solution:

$$C_i = (1 + f_1)(1 + f_2)(1 + f_3)\, C_P$$
$$= (1.030)(1.042)(1.047)(\$475{,}000) = \$533{,}800$$

Equipment Sizing Before equipment costs can be obtained, it is necessary to determine equipment size from material and energy balances. For preliminary estimates, rules of thumb may be used; but for definitive and detailed estimates, detailed equipment calculations must be made.

Example 3: Equipment Sizing and Costing Oil at 490,000 lb/h is to be heated from 100 to 170°F with 145,000 lb/h of kerosene initially at 390°F from another section of a plant. The oil enters at 20 psig and the kerosene at 25 psig. The physical properties are

Oil—0.85 sp gr, 3.5 cP at 135°F, 0.49 sp ht

Kerosene—0.82 sp gr, 0.45 cP, 0.61 sp ht

Estimate the cost of an all-carbon-steel exchanger in late 2004. Assume a counterflow shell-and-tube exchanger.

Solution:

Energy required to heat oil stream (490,000)(0.49)(170 – 100) = 16,807,000 Btu/h

$$\text{Exit kerosene temperature } T = 390 - \left(\frac{490{,}000}{145{,}000}\right)\left(\frac{0.49}{0.61}\right)(170 - 100)$$

$$= 200^\circ \text{F}$$

$$\text{LMTD} = \frac{220 - 100}{\ln 2.2} = 152^\circ F$$

Calculate the exchanger efficiency factor, F.

$$P = \frac{170 - 100}{390 - 100} = 0.241$$

$$R = \frac{390 - 200}{170 - 100} = 2.71$$

From Perry $F = 0.88$. Since the factor must be greater than 0.75, the exchanger is satisfactory. Therefore, $\Delta T = (F)(\text{LMTD}) = (0.88)(152) = 134^\circ$F.

Assume $U_D = 50$ Btu/(h·ft²·°F).

$$Q = U_D A\, \Delta T = 16{,}800{,}000 = (50)(A)(134)$$
$$A = 2510 \text{ ft}^2$$

Use the cost algorithm cited above.

$$C_B = \exp\,[8.821 - 0.30863 \ln A + 0.0681(\ln A)^2]$$
$$= \exp\,[8.821 - 0.30863(7.83) + 0.0681(61.3)] = \$39{,}300 \text{ base cost}$$

$F_D = 1.0$ F_M = for cs/cs material = 1.0

$F_P = 1.00$ since this exchange is operating below 4 bar

$$K = 1.218\,(\text{CE index 4th qtr 2004/CE index 1st qtr 2003}) = 1.218\left(\frac{463}{406}\right)$$
$$= 1.389$$

Therefore, $C_{HE} = KC_BF_DF_MF_P = (1.389)(39{,}300)(1.0)(1.0)(1.0) = \$54{,}600$.

Estimation of Fixed Capital Investment

Order-of-Magnitude Methods The **ratio method** will give the fixed capital investment per gross annual sales; however, most of these data are from the 1960s, and no recent data have been published. The ratio above is called the *capital ratio*, often used by financial analysts. The reciprocal of the capital ratio is the *turnover ratio* that for various businesses ranges from 4 to 0.3. The chemical industry has an average of about 0.4 to 0.5. The ratio method of obtaining fixed capital investment is rapid but suitable only for order-of-magnitude estimates.

The **exponential method** may be used to obtain a rapid capital cost for a plant based upon existing company data or from published sources such as those of D. S. Remer and L. H. Chai, *Chemical Engineering*,

April 1990, pp. 138–175. In the method known as the *seven-tenths rule*, the cost-capacity data for process plants may be correlated by a logarithmic plot similar to the six-tenths plot for equipment. Remer and Chai compiled exponents for a variety of processes and found that the exponents ranged from 0.6 to 0.8. When the data are used to obtain a capital cost for a different-size plant, the estimated capital must be for the same process.

The equation is

$$\text{Cost of plant B} = \text{cost of plant A}\left(\frac{\text{capacity of plant B}}{\text{capacity of plant A}}\right)^{0.7} \quad (9\text{-}5)$$

Cost indices may be used to correct costs for time changes.

Example 4: Seven-Tenths Rule A company is considering the manufacture of 150,000 tons annually of ethylene oxide by the direct oxidation of ethylene. According to Remer and Chai (1990), the cost capacity exponent for such a plant is 0.67. A subsidiary of the company built a 100,000-ton annual capacity plant for $70 million fixed capital investment in 1996. Using the seven-tenths rule, estimate the cost of the proposed new facility in the third quarter 2004.

Solution:

$$\text{Cost}_{150} = \text{Cost}_{100}\left(\frac{\text{Cap}_{150}}{\text{Cap}_{100}}\right)^{0.67}\left(\frac{\text{CE index 3Q 2004}}{\text{CE index 1996}}\right)$$

$$= (\$70{,}000{,}000)\left(\frac{150{,}000}{100{,}000}\right)^{0.67}\left(\frac{457.4}{381.7}\right) = \$110{,}000{,}000$$

Study Method The single-factor method begins with collecting the delivered cost of various items of equipment and applying one factor to obtain the battery-limits (BL) fixed capital (FC) investment or total capital investment as follows:

$$(C_{FC})_{BL} = f\,\Sigma(C_{EQ})_{DEL} \quad (9\text{-}6)$$

where $(C_{FC})_{BL}$ = battery-limits fixed capital investment or total capital investment
$(C_{EQ})_{DEL}$ = delivered equipment costs

The single factors include piping, automatic controls, insulation, painting, electrical work, engineering costs, etc. (Couper, 2003). Table 9-10 shows the Lang factors for various types of processing plants. The boundaries between the classifications are not clear-cut, and considerable judgment is required in the selection of the appropriate factors.

Preliminary Estimate Methods A refinement of the Lang factor method is the Hand method. The Hand factors are found in Table 9-11. Equipment is grouped in categories, such as heat exchangers and pumps, and then a factor is applied to each group to obtain the installed cost; finally the groups are summed to give the battery-limits installed cost. Wroth compiled a more detailed list of installation factors; a selection of these can be found in Table 9-12. The Lang and Hand methods start with purchased equipment costs whereas the Wroth method begins with delivered equipment costs, so delivery charges must be included in the Lang and Hand methods. At best the Lang and Hand methods will yield study quality estimates, and the Wroth method might yield a preliminary quality estimate.

Example 5: Fixed Capital Investment Using the Lang, Hand, and Wroth Methods The following is a list of the purchased equipment costs for a proposed processing unit:

Heat exchangers	$620,000
Distillation towers and internals	975,000
Receivers	320,000
Accumulator drum	125,000
Pumps and motors	220,000
Automatic controls	275,000
Miscellaneous equipment	150,000

Assume delivery charges are 5 percent of the purchased price. Estimate the fixed capital investment 2 years into the future, using the Lang, Hand, and Wroth methods. The inflation rates are 3.5 percent for the first year and 4.0 percent for the second.

Solution:

Equipment	Purchased equipment cost	Delivered equipment cost
Heat exchangers	$620,000	$651,000
Distillation towers, internals	975,000	1,024,000
Receivers	320,000	336,000
Accumulator drum	125,000	131,000
Pumps and motors	220,000	231,000
Automatic controls	275,000	289,000
Miscellaneous equipment	150,000	158,000
Total	$2,685,000	$2,820,000

Lang method: The Lang factor for a fluid processing unit starting with purchased equipment costs is 5.0. Therefore, fixed capital investment is $2,820,000 × 5.0 × 1.035 × 1.040 = $15,177,000.

Hand method: The Hand method begins with purchased equipment costs, and factors are applied from Table 9-11.

TABLE 9-10 Lang Factors

	Lang factors	
Type of plant	Fixed capital investment	Total capital investment
Solid processing	4.0	4.7
Solid-fluid processing	4.3	5.0
Fluid processing	5.0	6.0

Adapted from M. S. Peters, K. D. Timmerhaus, and R. West, *Plant Design and Economics for Chemical Engineers*, 5th ed., McGraw-Hill, New York, 2004.

TABLE 9-11 Hand Factors

Equipment type	Factor
Fractionating columns	4.0
Pressure vessels	4.0
Heat exchangers	3.5
Fired heaters	2.0
Pumps	4.0
Compressors	2.5
Instruments	4.0
Miscellaneous equipment	2.5

Adapted from W. E. Hand, *Petroleum Refiner*, September 1958, pp. 331–334.

TABLE 9-12 Selected Wroth Factors

Equipment	Factor
Blender	2.0
Blowers and fans	2.5
Centrifuge	2.0
Compressors	
Centrifugal (motor-driven)	2.0
Centrifugal (steam-driven, including turbine)	2.0
Reciprocating (steam and gas)	2.3
Reciprocating (motor-driven less motor)	2.3
Ejectors, vacuum	2.5
Furnaces (packaged units)	2.0
Heat exchangers	4.8
Instruments	4.1
Motors, electric	3.5
Pumps	
Centrifugal (motor-driven less motor)	7.0
Centrifugal (steam-driven including turbine)	6.5
Positive-displacement (less motor)	5.0
Reactors (factor as appropriate, equivalent-type equipment)	—
Refrigeration (packaged units)	2.5
Tanks	
Process	4.1
Storage	3.5
Fabricated and field-erected 50,000+ gal	2.0
Towers (columns)	4.0

Abstracted from W. F. Wroth, *Chemical Engineering*, October 17, 1960, p. 204.

Hand method:

Equipment	Purchased equipment cost	Hand factor	Purchased equipment installed cost
Heat exchangers	$620,000	3.5	$2,170,000
Distillation towers, internals	975,000	4.0	3,900,000
Receivers	320,000	2.5*	800,000
Accumulator drum	125,000	2.5*	313,000
Pumps and motors	220,000	4.0	880,000
Automatic controls	275,000	4.0	1,100,000
Miscellaneous	150,000	2.5	375,000
TOTAL	$2,685,000		$9,538,000

The asterisk on the receivers and accumulators indicates that if these vessels are pressure vessels, a factor of 4.0 should be used instead of 2.5. The total purchased equipment installed is $9,538,000 for non–pressure vessels and the delivered cost is $10,015,000. Therefore, the fixed capital investment installed would be $10,015,000 × 1.035 × 1.040 = $10,780,000. Using pressure vessels increases the total purchased equipment cost $667,000; therefore, the fixed capital investment for this case including inflation would be $10,780,000 × 1.05 × 1.035 × 1.04 = $11,534,000.

Wroth method:

Equipment	Delivered equipment cost	Wroth factor	Delivered equipment installed cost
Heat exchangers	$651,000	4.8	$3,125,000
Distillation towers, internals	1,024,000	4.0	4,096,000
Receivers	336,000	3.5	1,176,000
Accumulator drum	131,000	3.5	459,000
Pumps and motors	231,000	7.0	1,617,000
Automatic controls	289,000	4.1	1,185,000
Miscellaneous	158,000	4.0 (assumed)	632,000
TOTAL	$2,820,000		$12,290,000

The total delivered installed equipment cost is the fixed capital investment and, corrected for 2 years of inflation, will be $12,290,000 × 1.035 × 1.040 = $13,229,000.

Therefore, the summary of the fixed capital investment by the various methods is

Lang	$15,177,000
Hand	11,534,000
Wroth	13,229,000

Experience has shown that the fixed capital investment by the Lang method is generally higher than that of the other methods. Whatever figure is reported to management, it is advisable to state the potential accuracy of these methods.

Brown developed guidelines for the preparation of order-of-magnitude and study capital cost estimates based upon the Lang and Hand methods. Brown modified Lang and Hand methods for materials of construction, instrumentation, and location factors. He found that the modified Hand and Garrett module factor methods gave results within 3.5 percent.

Other multiple-factor methods that have been published in the past are those by C. E. Chilton, *Cost Estimation in the Process Industries*, McGraw-Hill, New York, 1960; M. S. Peters, K. D. Timmerhaus, and R. E. West, *Plant Design and Economics for Chemical Engineers*, 5th ed., McGraw-Hill, New York, 2003; C. A. Miller, *Chemical Engineering*, Sept. 13, 1965, pp. 226–236; and F. A. Holland, F. A. Watson, and V. K. Wilkinson, *Chem. Eng.*, Apr. 1, 1974, pp. 71–76. These methods produced preliminary quality estimates. Most companies have developed their own in-house multiple-factor methods for preliminary cost estimation.

Step-counting methods are based upon a number of processing steps or "functional units." The concept was first introduced by H. E. Wessel, *Chem. Eng.*, 1952, p. 209. Subsequently, R. D. Hill, *Petrol. Refin.*, **35**(8):106–110, August 1956; F. C. Zevnik and R. L. Buchanan, *Chem. Eng. Progress*, **59**(2):70–77, February 1963; and J. H. Taylor, *Eng. Process Econ.*, **2**:259–267, 1977, further developed the step-counting method.

A step or functional unit is a significant process step including all process equipment and ancillary equipment necessary for operating the unit. A functional unit may be a unit operation, unit process, or separation in which mass and energy are transferred. The sum of all functional units is the total fixed capital investment. Pumping and heat exchangers are considered as part of a functional unit. In-process storage is generally ignored except for raw materials, intermediates, or products. Difficulties are encountered in applying the method due to defining a step. This takes practice and experience. If equipment has been omitted from a step, the resulting estimate is seriously affected. These methods are reported to yield estimates of study quality or at best preliminary quality.

Definitive Estimate Methods Modular methods are an extension of the multiple-factor methods and have been proposed by several authors. One of the most comprehensive methods and one of the earliest was that of K. M. Guthrie, *Chem. Eng.*, **76:**114–142, Mar. 24, 1969. It began with equipment FOB equipment costs, and through the use of factors listed in Table 9-13, the module material cost was obtained. Labor for erection and setting equipment was added to the material cost as well as indirect costs for freight, insurance, engineering, and field expenses to give a total module cost. Such items as contingencies, contractors' fees, auxiliaries, site development land, and industrial buildings were added if applicable. Since any plant consists of equipment modules, these are summed to give the total fixed capital investment. Unfortunately, the factors and data are old but the concept is useful. Garrett (1989) developed a similar method based upon a variety of equipment modules, starting with purchased equipment costs obtained from plots and applying factors for materials of construction, instrumentation, and plant location. The method provides for all supporting and connecting equipment to make the equipment installation operational. See Table 9-14. T. R. Brown, *Hydrocarbon Processing*, October 2000, pp. 93–100, made modifications to the Garrett method.

Another method, called the *discipline method*, mentioned by L. R. Dysert, *Cost Eng.* **45**(6), June 6, 2003, is similar to the models of

TABLE 9-13 Guthrie Method Factors*

Details	Exchangers: Furnaces	Exchangers: Shell and tube	Exchangers: Air-cooled	Vessels: Vertical	Vessels: Horizontal	Pump and driver	Compressor and driver	Tanks
FOB equipment	1.00	1.00	1.00	1.00	1.00	1.00	1.00	1.00
Piping	0.18	0.46	0.18	0.61	0.42	0.30	0.21	
Concrete	0.10	0.05	0.02	0.10	0.06	0.04	0.12	
Steel		0.03		0.08				
Instruments	0.04	0.10	0.05	0.12	0.06	0.03	0.08	
Electrical	0.02	0.02	0.12	0.05	0.05	0.31	0.16	
Insulation		0.05		0.08	0.05	0.03	0.03	
Paint			0.01	0.01	0.01	0.01	0.01	
Total materials = *M*	1.34	1.71	1.38	2.05	1.65	1.72	1.61	1.20
Erection and setting (*L*)	0.30	0.63	0.38	0.95	0.59	0.70	0.58	0.13
χ, excluding site preparation and auxiliaries (*M* + *L*)	1.64	2.34	1.76	3.00	2.24	2.42	2.19	1.33
Freight, insurance, taxes, engineering, home office, construction		0.08		0.08	0.08	0.08	0.08	0.08
Overhead or field expense	0.60	0.95	0.70	1.12	0.92	0.97	0.97	
Total module factor	2.24	3.37	2.46	4.20	3.24	3.47	3.24	1.41

*From K. M. Guthrie, *Chem. Eng.*, **76,** 114–142 (Mar. 24, 1969). Based on FOB equipment cost = 100 (carbon steel).

TABLE 9-14 Selected Garrett Module Factors

Equipment type (carbon steel unless otherwise noted)	Module factor
Agitators: dual-bladed turbines/single-blade propellers	2.0
Agitated tanks	2.5
Air conditioning	1.46
Blender, ribbon	2.0
Blowers, centrifugal	2.5
Centrifuges: solid-bowl, screen-bowl, pusher, stainless steel	2.0
Columns: distillation, absorption, etc.	
Horizontal	3.05
Vertical	4.16
Compressors: low-, medium-, high-pressure	2.6 avg.
Coolers, quenchers	2.7
Crystallizers	2.6 avg.
Drives/motors	
Electric, for fans, compressors, pumps	1.5
Electric for other units	2.0
Gasoline	2.0
Turbine: gas and steam	3.5
Dryers	
Fluid bed, spray	2.7
Rotary	2.3
Dust collectors	
Bag filters	2.2
Cyclones, multiclones	3.0
Evaporators, single-effect stainless steel	
Falling film	2.3
Forced circulation	2.9
Fans	2.2
Filters	
Belt, rotary drum and leaf, tilting pan	2.4
Others	2.8
Furnaces	2.1
Heat exchangers	
Air-cooled	2.2
Double-pipe	1.8
Shell-and-tube	3.2
Mills	
Hammer	2.8
Ball, rod	2.3 avg.
Pumps	
Centrifugal	5.0
Reciprocating	3.3
Turbine	1.8
Reactors, jacketed, no agitator	
304 SS	1.8
Glass-lined	2.1
Mild steel	2.3
Vacuum equipment	2.2

SOURCE: Adapted from Garrett (1989).

TABLE 9-15 Code of Accounts

Category number	Direct capital cost account titles
010	Equipment items
020	Instrument items
030	Setting and testing equipment
040	Setting and testing instruments
050	Piling
060	Excavation
070	Foundations
080	Supports, platforms, and structures
090	Other building items
100	Fire protection and sprinklers
110	Piping
120	Ductwork
130	Electrical and wiring
140	Site preparation
150	Sewers, drains, and plumbing
160	Underground piping
170	Yards, roads, and fencing
180	Railroads
190	Insulation
200	Painting
210	Walls, masonry, roofs, and roofing
220	Spares
230	Lump-sum contracts
	Distributives
500	Site burden
510	Direct labor burden
530	Construction equipment, tools, and supplies
550	Rental and servicing construction equipment and tools
580	Premium wages and overtime—contractor
670	Temporary facilities
740	Cancellation charges
750	Abandoned design
760	Self-insured losses
790	Unvouchered liabilities
800	In-house engineering
810	Outside engineering
870	Undeveloped design allowances
880	Distributives transferred to expense
890	Contingencies—capital items
	Expense
900	Dismantling
910	Sales and use taxes
920	Repairs expense
930	Relocation and modification expense
940	Start-up relocation and modification expense
990	Contingencies

SOURCE: Private communication.

Guthrie and Garrett. It uses equipment factors to generate separate costs for each of the "disciplines" associated with the installation of equipment, such as installation labor, concrete, structural steel, and piping, to obtain direct field costs for each type of equipment, e.g., heat exchangers, towers, and reactors.

Modular methods, depending on the amount of detail provided, will yield preliminary quality estimates.

Detailed Estimate Method For estimates in the detailed category, a code of account needs to be used to prevent oversight of certain significant items in the capital cost. See Table 9-15. Each item in the code is estimated and provides the capital cost estimate; then this estimate serves for cost control during the construction phase of a project.

Comments on Significant Cost Items

Piping This cost includes the cost of the pipe, installation labor, valves, fittings, supports, and miscellaneous items necessary for complete installation of all pipes in the process. The accuracy of the estimates can be seriously in error by the improper application of estimating techniques to this component. Many pipe estimating methods are extant in the literature.

Two general methods have been used to estimate piping costs when detailed flow sheets are not available. One method is to use a percentage of the FOB equipment costs or a percentage of the fixed capital investment. Typical figures are 80 to 100 percent of the FOB equipment costs or 20 to 30 percent of the fixed capital investment. This method is used for preliminary estimates. Another group of methods such as the Dickson "*N*" method (R. A. Dickson, *Chem. Eng.*, **57**:123–135, Nov. 1947), estimating by weight, estimating by cost per joint, etc., requires a detailed piping takeoff from either PID or piping drawings with piping specifications, material costs, labor expenses, etc. These methods are used for definitive or detailed estimates where accuracy of 10 to 15 percent is required. The takeoff methods must be employed with great care and accuracy by an experienced engineer. A detailed breakdown by plant type for process piping costs is presented in Peters et al. (2003) and in *Perry's Chemical Engineers' Handbook*, 6th ed., 1984.

Electrical This item consists of transformers, wiring, switching gear, as well as instrumentation and control wiring. The installed costs of the electrical items may be estimated as 20 to 40 percent of the delivered equipment costs or 5 to 10 percent of the fixed capital investment for preliminary estimates. As with piping estimation, the process design must be well along toward completion before detailed electrical takeoffs can be made.

Buildings and Structures The cost of the erection of buildings and structures in a chemical process plant as well as the plumbing, heating and ventilation, and miscellaneous building service items may be estimated as 20 to 50 percent of delivered equipment costs or as 10 to 20 percent of the fixed capital investment for a preliminary estimate.

Yards, Railroad Sidings, Roads, etc. This investment includes roads, railroad spurs, docks, and fences. A reasonable figure for preliminary estimates is 15 to 20 percent of the FOB equipment cost or 3 to 7 percent of the fixed capital investment for a preliminary estimate.

Service Facilities For a process plant, utility services such as steam, water, electric power, fuel, compressed air, shop facilities, and a cafeteria require capital expenditures. The cost of these facilities lumped together may be 10 to 20 percent of the fixed capital investment for a preliminary estimate. (*Note:* Buildings, yards, and service facilities must be well defined to obtain a definitive or detailed estimate.)

Environmental Control and Waste Disposal These items are treated as a separate expenditure and are difficult to estimate due to the variety and complexity of the process requirements. Pollution control equipment is generally included as part of the process design. Couper (2003) and Peters et al. (2003) mention that at present there are no general guidelines for estimating these expenditures.

Computerized Cost Estimation With the advent of powerful personal computers (PCs) and software packages, capital cost estimates advanced from large mainframe computers to the PCs. The reasons for using computer cost estimation and economic evaluation packages are time saved on repetitive calculations and reduction in mathematical errors. Numerous computer simulation software packages have been developed over the past two decades. Examples of such software are those produced by ASPEN, ICARUS, CHEMCAD, SUPERPRO, PRO II, HYSYS, etc.; but most do not contain cost estimation software packages. ICARUS developed a PC cost estimation and economic evaluation package called Questimate. This system built a cost estimate from design and equipment cost modules, bulk items, site construction, piping and ductwork, buildings, electrical equipment, instruments, etc., developing worker-hours for engineering and fieldwork costs. This process is similar to quantity takeoff methods to which unit costs are applied. A code of accounts is also provided.

ASPEN acquired ICARUS in 2000 and developed Process Evaluator based on Questimate that is used for conceptual design, known as front-end loading (FEL). More information on FEL and value-improving process (VIP) is found later in Sec. 9. Basic and detailed estimates are coupled with a business decision framework in ASPENTECH ICARUS 2000.

EstPro is a process plant cost estimation package for conceptual cost estimation for conceptual design only. It may be obtained from Gulf Publishing, Houston, Tex.

Many companies have developed their own factored estimates using computer spreadsheets based upon their in-house experience and cost database information that they have built from company project history. For detailed estimates, the job is outsourced to design-construction companies that have the staff to perform those estimates.

Whatever package is used, it is recommended that computer-generated costs be spot-checked for reasonable results using a handheld calculator since errors do occur. Some commercial software companies will develop cost estimation databases in cooperation with a company for site-specific costs.

Contingency This is a provision for unforeseen events that experience has demonstrated are likely to occur. Contingencies are of two types: process and project contingency. In the former, there are uncertainties in

Equipment and performance
Integration of old and new process steps
Scaling up to a large-scale plant size
Accurate definition of certain process parameters, such as severity of process conditions, number of recycles, process blocks and equipment, multiphase streams, and unusual separations

No matter how much time and effort are spent preparing estimates, there is a chance of errors occurring due to

Engineering errors and omissions
Cost and labor rate changes
Construction problems
Estimating inaccuracies
Miscellaneous "unforeseens"
Weather-related problems
Strikes by fabricators, transportation, and construction personnel

For preliminary estimates, a 15 to 20 percent project contingency should be applied if the process information is firm. As the quality of the estimate moves to definitive and detailed, the contingency value may be lowered to 10 to 15 percent and 5 to 10 percent, respectively. Experience has shown that the smaller the dollar value of the project, the higher the contingency should be.

Offsite Capital These facilities include all structures, equipment, and services that do not enter into the manufacture of a product but are important to the functioning of the plant. Such capital items might be steam-generating and electrical-generating and distribution facilities, well-water cooling tower, and pumping stations for water distribution, etc. Service capital might be auxiliary buildings, such as warehouses, service roads, railroad spurs, material storage, fire protection equipment, and security systems. For estimating purposes, the following percentages of the fixed capital investment might be used:

Small modification of offsites, 1 to 5 percent
Restructuring of offsites, 5 to 15 percent
Major expansion of offsites, 15 to 45 percent
Grass-roots plants, 45 to 150 percent

Allocated Capital This is capital that is shared due to its proportionate share use in a new facility. Such items include intermediate chemicals, utilities, services and sales, administration, research, and engineering overhead.

Working Capital Working capital is the funds necessary to conduct day-to-day company business. These are funds required to purchase raw materials, supplies, etc. It is continuously liquidated and rejuvenated from the sale of products or services. If an adequate amount of working capital is available, management has the necessary flexibility to cover expenses in case of strikes, delays, fires, etc. Several methods are available for estimating an adequate amount of working capital. They may be broadly classified into percentage and inventory methods. The percentage methods are satisfactory for study and preliminary capital estimates. The percentage methods are of two types: percentage based on capital investment and percentage based upon sales. In the former method, 15 to 25 percent of the total capital investment may be sufficient for preliminary estimates. In the case of certain specialty chemicals where the raw materials are expensive, it is perhaps better to use the percentage of sales method. Such chemicals as flavors, fragrances, perfumes, etc., are in this category. Experience has shown that 15 to 45 percent of sales has been used with 30 to 35 percent being a reasonable average value.

Start-up Expenses Start-up expenses are defined as the total costs directly related to bringing a new manufacturing facility onstream. Start-up time is the time span between the end of construction and the beginning of normal operation. Normal operation is operation at a certain percentage of design capacity or a specified number of days of continuous operation or the ability to make product of a specified purity. Start-up costs are part of the total capital investment and include labor, materials, and overhead for design modifications or changes due to errors on the part of engineering, contractors, costs of tests, final alterations and adjustments. These items cannot be included as contingency because it is known that such work will be necessary before the project is completed. Experience has shown that start-up costs are a percentage of the battery-limits fixed capital investment of the order on average of 3 percent.

Depending on the tax laws in effect, not all start-up costs can be expensed and a portion must be capitalized. Start-up costs can reduce the after-tax earnings during the early years of a project because of a delay in the start-up of production causing a loss of earnings. Construction changes are items of capital cost, and production start-up costs are expensed as an operating expense.

Other Capital Items Paid-up royalties and licenses are considered part of the capital investment since these are replacements for capital to perform process research and development. The initial catalyst and chemical charge, especially for noble metal catalysts and/or in electrolytic processes, is a large amount. These materials are considered to have a life of 1 year. If funds must be borrowed for a new facility, then the interest on borrowed funds during the construction period is capitalized; otherwise, the interest is part of the operating expense.

MANUFACTURING-OPERATING EXPENSES

The estimation of manufacturing expenses has received less attention in the open literature than the estimation of capital requirements. Operating expenses are estimated from proprietary company files. In this section, methods for estimating the elements that constitute operating expenses are presented. Operating expenses consist of the expense of manufacturing a product, packaging and shipping, as well as general overhead expense. These are described later in this section. Figure 9-6 shows an example of a typical manufacturing expense sheet.

RAW MATERIAL EXPENSE

Estimates of the amount of raw material consumed can be obtained from the process material balance. Normally, the raw material expense is the largest expense item in the manufacture of a product. Since yields in a chemical reaction determine the quantity of raw materials consumed, assumed yields may be used to obtain approximate exploratory estimates if possible ranges are given. The prices of the raw materials are published in various trade journals that list material according to form, grade, method of delivery, unit of measure, and cost per unit. The *Chemical Marketing Reporter* is a typical source of these prices. The prices are generally higher than quotations from suppliers, and these latter should be used whenever possible. It may be possible for a company to negotiate the price of a raw material based upon large-quantity use on a long-term basis. With the amount of material used from the material balance and the price of the raw material, the following information can be obtained: annual material consumption, annual material expense, as well as the consumption and expense per unit of product.

Occasionally, by-products may be produced, and if there is a market for these materials, a credit can be given. By-products are treated in the same manner as raw materials and are entered into the manufacturing expense sheet as a credit. If by-products are intermediates for which no market exists, they may be credited to downstream or subsequent operations at a value equivalent to their value as a replacement, or no credit may be obtained.

DIRECT EXPENSES

These are the expenses that are directly associated with the manufacture of a product, e.g., utilities, labor, and maintenance.

Utilities The utility requirements are obtained from the material and energy balances. Utilities include steam, electricity, cooling water, fuel, compressed air, and refrigeration. The current utility prices can be obtained from company plant accounting or from the plant utility supervisor. This person might be able to provide information concerning rate prices for the near future. As requirements increase, the unit cost declines. If large incremental amounts are required, e.g., electricity, it may be necessary to tie the company's utility line to a local utility as a floating source.

With the current energy demands increasing, the unit costs of all utilities are increasing. Any prices quoted need to be reviewed periodically to determine their effect on plant operations. A company utility supervisor is a good source of future price trends. Unfortunately, there are no shortcuts for estimating and projecting utility prices. Utilities are the third largest expense item in the manufacture of a product, behind raw materials and labor.

Operating Labor The most reliable method for estimating labor requirements is to prepare a table of shift, weekend, and vacation coverage. For round-the-clock operation of a continuous process, one operator per shift requires 4.2 operators, if it is assumed that 21 shifts cover the operation and each operator works five, 8-h shifts per week. For batch or semicontinuous operation, it is advisable to prepare a labor table, listing the number of tasks and the number of operators required per task, paying particular attention to primary processing steps such as filtration and distillation that may have several items of equipment per step.

Labor rates may be obtained from the union contract or from a company labor relation supervisor. This person will know the current labor rates and any potential labor rate increases in the near future. One should not forget shift differential and overtime charges. Once the number of operators per shift has been established, the annual labor expense and unit expense may be estimated. Wessel (*Chem. Eng.,* **59:**209–210, July 1952) developed a method for estimating labor requirements for various types of chemical processes in the United States. The equation is applicable for a production rate of 2 to 2000 tons/day (2000 lb/ton).

$$\log Y = -0.783 \log X + 1.252 + B \qquad (9\text{-}7)$$

where Y = operating labor, operator h/ton per processing step
X = plant capacity, tons/day
B = constant depending upon type of process
+ 0.132 (for batch operations that have minimum labor requirements)
+ 0 (for operations with average labor requirements)
− 0.167 (for a well-instrumented continuous process)

A processing step is one in which a unit operation occurs; e.g., a filtration step might consist of a feed (precoat) tank, pump, filter, and receiver so a processing step may have several items of equipment. By using a flow sheet, the number of processing steps may be counted. The Wessel equation does not take into account changes in labor productivity, but this information can be obtained from each issue of *Chemical Engineering.* Labor productivity varies widely in various sections of this country but even more widely in foreign countries.

Ulrich (1984) developed a table for estimating labor requirements from flow sheets and drawings of the process. Consideration is given to the type and arrangement of equipment, multiplicity of units, and amount of process control equipment. This method is easier to use than the Wessel method and has been updated in a new edition of the original text.

Supervision The approximate expense for supervision of operations depends on process complexity, but 15 to 30 percent of the operating labor expense is reasonable.

Payroll Charges This item includes workers' compensation, social security premiums, unemployment taxes, paid vacations, holidays, and some part of health and dental insurance premiums. The figure has steadily declined from 1980 and now is 30 to 40 percent of operating labor plus supervision expenses.

Maintenance The maintenance expense consists of two components, namely, materials and labor, approximately 60 and 40 percent, respectively. Company records are the best information sources, however, a value of 6 to 10 percent of the fixed capital investment is a reasonable figure. Processes with a large amount of rotating equipment or that operate at extremes of temperature and/or pressure have higher maintenance requirements.

Miscellaneous Direct Expenses These items include operating supplies, clothing and laundry, laboratory expenses, royalties, environmental control expenses, etc.

Item	Basis	Percentage
Operating supplies	Operating labor	5–7
Clothing and laundry	Operating labor	10–15
Laboratory expenses	Operating labor	10–20
Royalties and patents	Sales	1–5

Environmental Control Expense Wastes from manufacturing operations must be disposed of in an environmentally acceptable manner. This direct expense is borne by each manufacturing department. Some companies have their own disposal facilities, or they may contract with a firm that handles the disposal operation. However the wastes are handled, there is an expense. Published data are found in the open literature, some of which have been published by Couper (2003).

TOTAL OPERATING EXPENSES

PRODUCT:	PLASTICIZER X			
TOTAL SALES ($/YR):	7200000			
RATED CAPACITY (MM LBS/YR):	12			
LOCATION:				
FIXED CAPITAL INVESTMENT:	800000			
LAND	25000			
WORKING CAPITAL	120000			
OPERATING HOURS (HRS/YR):				
DATE:				
BY:				
RAW MATERIALS:				
MATERIAL	UNIT	ANNUAL QUANTITY	$/UNIT	$/YEAR
A AND B	LB	12000000	.23	2760000
				0
				0
				0
GROSS MATERIAL EXPENSE				2760000
BY-PRODUCTS:	UNIT	ANNUAL QUANTITY	$/UNIT	$/YEAR
				0
				0
BY-PRODUCT CREDIT				0
NET MATERIAL EXPENSE				2760000
DIRECT EXPENSES:				
	UNIT	ANNUAL QUANTITY	$/UNIT	$/YEAR
UTILITIES:				
steam, low pressure				0
steam, medium pressure				0
steam, high pressure	LB	60000000	.003	180000
GROSS STEAM EXPENSE				180000
STEAM CREDIT				0
NET STEAM EXPENSE				180000
electricity	KWH	3000000	.035	105000
cooling water	GALLONS	72000000	.000045	3240
fuel gas				0
other:				0
city water	GALLONS	360000000	.0002	72000
TOTAL UTILITIES COST				540240
LABOR:				
men per shift	4			
annual labor rate per shift	25000			25000
TOTAL LABOR COSTS				100000
SUPERVISION:				
% total of labor expense				0
SUPERVISION EXPENSE=				18000
PAYROLL CHARGES, FRINGE BENEFITS:				
% total of labor expense	40			
PAYROLL EXPENSE				47200
MAINTENANCE				
% of fixed capital investment	8			
MAINTENANCE EXPENSE				64000
SUPPLIES:				
% of operating labor				0
SUPPLIES EXPENSE				1800
LABORATORY:				
laboratory hours per year	900			
cost per hour	30			
TOTAL LABORATORY EXPENSE				27000

FIG. 9-6 Total operating expense sheet.

ROYALTIES			
WASTE DISPOSAL:			0
tons per year			
waste charge per ton			0
WASTE DISPOSAL EXPENSE			0
OTHER:			
laundry	6000		6000
TOTAL DIRECT EXPENSE			804240
TOTAL DIRECT + NET MATERIAL COSTS			3564240
INDIRECT EXPENSES:			
DEPRECIATION			
% of fixed capital investment	100		
life of project (yrs)	7		
DEPRECIATION			114000
PLANT INDIRECT EXPENSES			
% of fixed capital investment	5		
PLANT INDIRECT EXPENSES			40000
TOTAL INDIRECT EXPENSES			154000
TOTAL MANUFACTURING EXPENSE:			3718240
PACKAGING, SHIPPING EXPENSE			
rated capacity per	12000000		
dollars per unit	.005		
PACKAGING AND SHIPPING EXPENSE			60000
TOTAL PRODUCTION EXPENSE			3778240
GENERAL OVERHEAD EXPENSES			
percent of annual sales		5	
GENERAL OVERHEAD EXPENSES			360000
TOTAL OPERATING EXPENSE			4138240

FIG. 9-6 *(Continued)*

INDIRECT EXPENSES

These indirect expenses consist of two major items; depreciation and plant indirect expenses.

Depreciation The Internal Revenue Service allows a deduction for the "exhaustion, wear and tear and normal obsolescence of equipment used in the trade or business." (This topic is treated more fully later in this section.) Briefly, for manufacturing expense estimates, straight-line depreciation is used, and accelerated methods are employed for cash flow analysis and profitability calculations.

Plant Indirect Expenses These expenses cover a wide range of items such as property taxes, personal and property liability insurance premiums, fire protection, plant safety and security, maintenance of plant roads, yards and docks, plant personnel staff, and cafeteria expenses (if one is available). A quick estimate of these expenses based upon company records is on the order of 2 to 4 percent of the fixed capital investment. Hackney presented a method for estimating these expenses based upon a capital investment factor, and a labor factor, but the result is high.

TOTAL MANUFACTURING EXPENSE

The total manufacturing expense for a product is the sum of the raw materials and direct and indirect expenses.

PACKAGING AND SHIPPING EXPENSES

The packaging expense depends on how the product is sold. The package may vary from small containers to fiberpacks to leverpacks, or the product may be shipped via tank truck, tank car, or pipeline. Each product must be considered and the expense of the container included on a case-by-case basis. The shipping expense includes the in-plant movement to warehousing facilities. Product delivery expenses are difficult to estimate because products are shipped in various amounts to numerous destinations. Often these expenses come under the heading of freight allowed in the sale of a product.

TOTAL PRODUCT EXPENSE

The sum of the total manufacturing expense and the packaging and in-plant shipping expense is the total product expense.

GENERAL OVERHEAD EXPENSE

This expense is often separated from the manufacturing expenses. It includes the expense of maintaining sales offices throughout the country, staff engineering departments, and research and development facilities and administrative offices. All manufacturing departments are expected to share in these expenses so an appropriate charge is made for each product varying between 6 and 15 percent of the product's annual revenue. The wide range in percentage will vary depending on the amount of customer service required due to the nature of the product.

TOTAL OPERATING EXPENSE

The sum of the total product expense and the general overhead expense is the total operating expense. This item ultimately becomes part of the operating expense on the income statement.

RAPID MANUFACTURING EXPENSE ESTIMATION

Holland et al. (1953) developed an expression for estimating annual manufacturing expenses for production rates other than the base case based upon fixed capital investment, labor requirements, and utility expense.

$$A_1 = mC_{\text{fci}} + nc_L N_1 + pU_1 \tag{9-8}$$

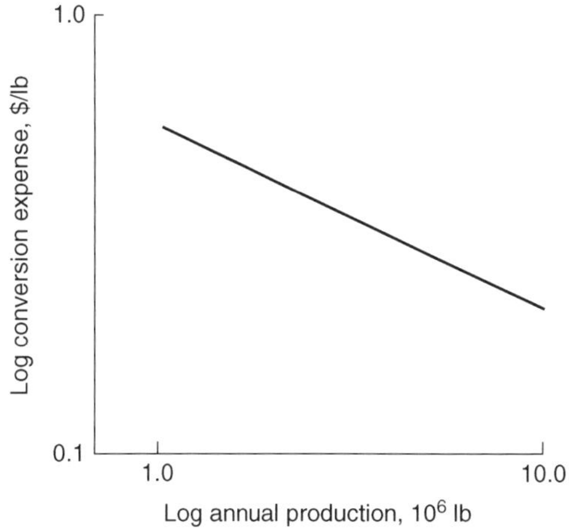

FIG. 9-7 Annual conversion expense as a function of production rate.

where C_{fci} = fixed capital investment, \$
C_L = cost of labor, \$ per operator per shift
N_1 = annual labor requirements, operators/shift/year at rate 1
U_1 = annual utility expenses at production rate 1
A_1 = annual conversion expense at rate 1
m, n, p = constants obtained from company records in consistent units

Equation (9-8) can be modified to include raw materials by adding a term qM_1, where q = a constant and M_1 = annual raw material expense at rate 1.

SCALE-UP OF MANUFACTURING EXPENSES

If it is desired to estimate the annual manufacturing expense at some rate other than a base case, the following modification may be made:

TABLE 9-16 Typical Labor Requirements for Various Equipment

Equipment	Laborers per unit per shift
Blowers and compressor	0.1–0.2
Centrifuge	0.25–0.50
Crystallizer, mechanical	0.16
Dryers	
Rotary	0.5
Spray	1.0
Tray	0.5
Evaporator	0.25
Filters	
Vacuum	0.125–0.25
Plate and frame	1.0
Rotary and belt	0.1
Heat exchangers	0.1
Process vessels, towers (including auxiliary pumps and exchangers)	0.2–0.5
Reactors	
Batch	1.0
Continuous	0.5

Adapted from G. D. Ulrich, *A Guide to Chemical Engineering Process Design and Economics*, Wiley, New York, 1984.

$$A_2 = mC_{fci}\left(\frac{R_2}{R_1}\right)^{0.7} + nC_LN_1\left(\frac{R_2}{R_1}\right)^{0.25} + pU_1\left(\frac{R_2}{R_1}\right) + qM_1\left(\frac{R_2}{R_1}\right) \quad (9\text{-}9)$$

where A_2 = annual manufacturing expense at production rate 2
R_1 = production rate 1
R_2 = production rate 2

Equation (9-9) may also be used to calculate data for a plot of manufacturing expense as a function of annual production rate, as shown in Fig. 9-7. Plots of these data show that the manufacturing expense per unit of production decreases with increasing plant size. The first term in Eq. (9-9) reflects the increase in the capital investment by using the 0.7 power for variations in production rates. Labor varies as the 0.25 power for continuous operations based upon experience. Utilities and raw materials are essentially in direct proportion to the amount of product manufactured, so the exponent of these terms is unity.

FACTORS THAT AFFECT PROFITABILITY

DEPRECIATION

According to the Internal Revenue Service (IRS), *depreciation* is defined as an allowance for the decrease in value of a property over a period of time due to wear and tear, deterioration, and normal obsolescence. The intent is to recover the cost of an asset over a period of time. It begins when a property is placed in a business or trade for the production of income and ends when the asset is retired from service or when the cost of the asset is fully recovered, whichever comes first. Depreciation and taxes are irrevocably tied together. It is essential to be aware of the latest tax law changes because the rules governing depreciation will probably change. Over the past 70 years, there have been many changes in the tax laws of which depreciation is a major component. Couper (2003) discussed the history and development of depreciation accounting. Accelerated depreciation was introduced in the early 1950s to stimulate investment and the economy. It allowed greater depreciation rates in the early years of a project when markets were not well established, manufacturing facilities were coming onstream, and expenses were high due to bringing the facility up to design capacity.

The current methods for determining annual depreciation charges are the *straight-line depreciation* and the *Modified Accelerated Cost Recovery System (MACRS)*. In the straight-line method, the cost of an asset is distributed over its expected useful life such that the annual charge is

$$D = \frac{I + S}{n} \quad \$ \text{ per year} \quad (9\text{-}10)$$

where D = annual depreciation charge
I = investment
n = number of years
S = salvage value

The MACRS went into effect in January 1987 (Couper, 2003) with six asset recovery periods: 3, 5, 7, 10, 15, and 20 years. It is based upon the declining-balance method. The equation for the declining-balance method is

$$V_e = V_i\,(1 - f) \quad (9\text{-}11)$$

where V_i = value of asset at beginning of year
V_e = value of asset at end of year
f = declining-balance factor

For 150 percent declining balance f = 1.5, and for 200 percent f = 2.0. These factors are applied to the previous year's remaining balance. It is evident that the declining-balance method will not recover the asset that the IRS permits. Therefore, a combination of the declining-balance and straight-line methods forms the basis for the MACRS.

Class lives for selected industries are found in Couper (2003), but most chemical processing equipment falls in the 5-year category and petroleum processing equipment in the 7-year category. For those assets with class lives less than 10 years, a 200 percent declining-balance

TABLE 9-17 Depreciation Class Lives and MACRS Recovery Periods

Asset class	Description of asset	Class life, yr	MACRS recovery period, yr
00.12	Information systems	6	5
00.4	Industrial steam and electric generation and/or distribution systems	22	15
13.3	Petroleum refining	16	10
20.3	Manufacture of vegetable oils and vegetable oil products	18	10
20.5	Manufacture of food and beverages	4	3
22.4	Manufacture of textile yams	8	5
22.5	Manufacture of nonwoven fabrics	10	7
26.1	Manufacture of pulp and paper	13	7
28.0	Manufacture of chemicals and allied products	9.5	5
30.1	Manufacture of rubber products	14	7
30.2	Manufacture of finished plastic products	11	7
32.1	Manufacture of glass products	14	7
32.2	Manufacture of cement	20	15
32.3	Manufacture of other stone and clay products	15	7
33.2	Manufacture of primary nonferrous metals	14	7
32.4	Manufacture of primary steel mill products	15	7
49.223	Substitute natural gas-coal gasification	18	10
49.25	Liquefied natural gas plant	22	15

SOURCE: "How to Depreciate Property," Publication 946, Internal Revenue Service, U.S. Department of Treasury, Washington, 1999.

method with a switch to straight-line in the later years is used. The IRS adopted a half-year convention for both depreciation methods. Under this convention, a property placed in service is considered to be only one-half year irrespective of when during the year the property was placed in service. Table 9-17 is a listing of the class lives, and Table 9-18 contains factors with the half-year convention for both the MACRS and straight-line methods.

Depreciation is entered as an indirect expense on the manufacturing expense sheet based upon the straight-line method. However, when one is determining the after-tax cash flow, straight-line depreciation is removed from the manufacturing expense and the MACRS depreciation is entered. This is illustrated under the section on cash flow.

There are certain terms that apply to depreciation:

- *Depreciation reserve* is the accumulated depreciation at a specific time.
- *Book value* is the original investment minus the accumulated depreciation.
- *Service life* is the time period during which an asset is in service and is economically feasible.
- *Salvage value* is the net amount of money obtained from the sale of a used property over and above any charges involved in the removal and sale of the property.
- *Scrap value* implies that the asset has no further useful life and is sold for the amount of scrap material in it.
- *Economic life* is the most likely period of successful operation before a need arises for subsequent investment in additional equipment as the result of product or process obsolescence or equipment due to wear and tear.

DEPLETION

Depletion is concerned with the diminution of natural resources. Generally depletion does not enter into process economic studies. Rules for determining the amount of depletion are found in the IRS Publication 535.

AMORTIZATION

Amortization is the ratable deduction for the cost of an intangible property over its useful life, perhaps a 15-year life, via straight-line calculations. An example of an intangible property is a franchise, patent, trademark, etc. Two IRS publications, Form 4562 and Publication 535 (1999), established the regulations regarding amortization.

TAXES

Corporations pay an income tax based upon gross earnings, as shown in Table 9-19. Most major corporations pay the federal tax rate of 34 percent on their annual gross earnings. In addition, some states have a stepwise corporate income tax rate. State income tax is deductible as an expense item before the calculation of the federal tax. If T_s is the incremental tax rate and T_f is the incremental federal tax, both expressed as decimals, then the combined incremental rate T_c is

TABLE 9-18 Depreciation Rates for Straight-Line and MACRS Methods

	Straight-line half-year convention							MACRS* half-year convention					
Year	3	5	7	10	15	20	Year	3	5	7	10	15	20
1	16.67%	10.00%	7.14%	5.0%	3.33%	2.5%	1	33.33%	20.00%	14.29%	10.00%	5.00%	3.750%
2	33.33	20.00	14.29	10.0	6.67	5.0	2	44.45	32.00	24.49	18.00	9.50	7.219
3	33.33	20.00	14.29	10.0	6.67	5.0	3	14.81	19.20	17.49	14.40	8.55	6.677
4	16.67	20.00	14.28	10.0	6.67	5.0	4	7.41	11.52	12.49	11.52	7.70	6.177
5		20.00	14.29	10.0	6.67	5.0	5		11.52	8.93	9.22	6.93	5.713
6		10.00	14.28	10.0	6.67	5.0	6		5.76	8.92	7.37	6.23	5.285
7			14.29	10.0	6.67	5.0	7			8.93	6.55	5.90	4.888
8			7.14	10.0	6.66	5.0	8			4.46	6.55	5.90	4.522
9				10.0	6.67	5.0	9				6.56	5.91	4.462
10				10.0	6.66	5.0	10				6.55	5.90	4.461
11				5.0	6.67	5.0	11				3.28	5.91	4.462
12					6.66	5.0	12					5.90	4.461
13					6.67	5.0	13					5.91	4.462
14					6.66	5.0	14					5.90	4.461
15					6.67	5.0	15					5.91	4.462
16					3.33	5.0	16					2.95	4.461
17						5.0	17						4.462
18						5.0	18						4.461
19						5.0	19						4.462
20						5.0	20						4.461
21						2.5	21						2.231

*General depreciation system. Declining-balance switching to straight-line. Recovery periods 3, 5, 7, 10, 15, and 20 years.

SOURCE: "How to Depreciate Property," Publication 946, Internal Revenue Service, U.S. Department of Treasury, Washington, 1999.

TABLE 9-19 Corporate Federal Income Tax Rates

Taxable income	Tax rate, %
Annual gross earnings less than $50,000	15
Annual gross earnings greater than $50,000 but not over $75,000	25
Annual gross earnings greater than $75,000 plus 5% of gross earnings over $100,000 or $11,750, whichever is greater	34
Corporations with gross earnings of at least $335,000 pay a flat rate of 34%	

SOURCE: U.S. Corporate Income Tax Return, Form 1120, Internal Revenue Service, U. S. Department of Treasury, Washington, 1999.

$$T_c = T_s + (1 - T_s)T_f \tag{9-12}$$

If the federal rate is 34 percent and the state rate is 7 percent, then the combined rate is

$$T_c = 0.07 + (1 - 0.07)(0.34) = 0.39$$

Therefore, the combined tax rate is 39 percent.

TIME VALUE OF MONEY

In business, money is either borrowed or loaned. If money is loaned, there is the risk that it may not be repaid. From the lender's standpoint, the funds could have been invested somewhere else and made a profit; therefore, the interest charged for the loan is compensation for the forgone profit. The borrower may look upon this interest as the cost of renting money. The amount of interest charged depends on the scarcity of money, the size of the loan, the length of the loan period, the risk that the lender feels that the loan may not be repaid, and the prevailing economic conditions. Engineers involved in the presentation and/or the evaluation of an investment of funds in a venture, therefore, need to understand the time value of money and how it is applied in the evaluation of projects.

The amount of the loan is called the *principal P*. The longer the time for which the money is loaned, the greater the total amount of interest paid. The *future amount of the money F* is greater than *the principal or present worth P*. The relationship between *F* and *P* depends upon the type of interest used. Table 9-20 is a summary of the nomenclature used in time value of money calculations.

Simple Interest The relationship between *F* and *P* is $F = P(1 + in)$. The interest is charged on the original loan and not on the unpaid balance (Couper and Rader, 1986). The interest is paid at the end of each time interval. Although the simple interest concept still exists, it is seldom used in business.

Discrete Compound Interest In financial transactions, loans or deposits are made using compound interest. The interest is not withdrawn but is added to the principal for that time period. In the next time period, the interest is calculated upon the principal plus the interest from the preceding time period. This process illustrates compound interest. In equation format,

$$\text{Year 1:} \quad P + Pi = P(1 + i) = F_1$$

$$\text{Year 2:} \quad P + Pi(1 + i) = P(1 + i)^2 = F_2 \tag{9-13}$$

$$\text{Year } n\text{:} \quad P(1 + i)^n = F$$

TABLE 9-20 Interest Nomenclature

Symbol	Definition
F	Future sum Future value Future worth Future amount
P	Principal Present worth Present value Present amount
A	End of period payment in a uniform series

An interest rate quoted on an annual basis is called *nominal interest*. However, interest may be payable on a semiannual, quarterly, monthly, or daily basis. To determine the amount compounded, the following equation applies:

$$F = P\left(1 + \frac{i}{m}\right)^{mn} \tag{9-14}$$

where m = number of interest periods per year
n = number of years
i = nominal interest

Interest calculated for a given time period is known as *discrete compound interest*, with *discrete* referring to a discrete time period. Table 9-21 contains 5 and 6 percent discrete interest factors.

Examples of the use of discrete factors for various applications are found in Table 9-22, assuming that the present time is when the first funds are expended.

Continuous Compound Interest In some companies, namely, petroleum, petrochemical, and chemical companies, money transactions occur hourly or daily, or essentially continuously. The receipts from sales and services are invested immediately upon receipt. The interest on this cash flow is continuously compounded. To use continuous compounding when evaluating projects or investments, one assumes that cash flows continuously.

In continuous compounding, the year is divided into an infinite number of periods. Mathematically, the limit of the interest term is

$$\lim_{n \to \infty}\left(1 + \frac{r}{m}\right)^{mn} = e^{rn} \tag{9-15}$$

where n = number of years
m = number of interest periods per year
r = nominal interest rate
e = base for naperian logarithms

The numerical difference between discrete compound interest and continuous compound interest is small, but when large sums of money are involved, the difference may be significant. Table 9-23 is an abbreviated continuous interest table, assuming that time zero is when start-up occurs. A summary of the equations for discrete compound and continuous compound interest is found in Table 9-24.

Compounding-Discounting When money is moved forward in time from the present to a future time, the process is called *compounding*. The effect of compounding is that the total amount of money increases with time due to interest. *Discounting* is the reverse process, i.e., a sum of money moved backward in time. Figure 9-8 is a

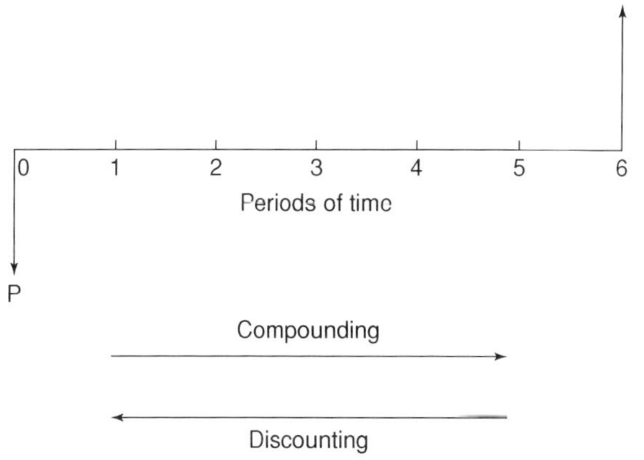

FIG. 9-8 Compounding-discounting diagram.

TABLE 9-21 Discrete Compound Interest Factors*

	Single payment		Uniform annual series				Single payment		Uniform annual series				
	Compound-amount factor	Present-worth factor	Sinking-fund factor	Capital-recovery factor	Compound-amount factor	Present-worth factor	Compound-amount factor	Present-worth factor	Sinking-fund factor	Capital-recovery factor	Compound-amount factor	Present-worth factor	
n	Given P, to find F $(1+i)^n$	Given F, to find P $\frac{1}{(1+i)^n}$	Given F, to find A $\frac{i}{(1+i)^n-1}$	Given P, to find A $\frac{i(1+i)^n}{(1+i)^n-1}$	Given A, to find F $\frac{(1+i)^n-1}{i}$	Given A, to find P $\frac{(1+i)^n-1}{i(1+i)^n}$	Given P, to find F $(1+i)^n$	Given F, to find P $\frac{1}{(1+i)^n}$	Given F, to find A $\frac{i}{(1+i)^n-1}$	Given P, to find A $\frac{i(1+i)^n}{(1+i)^n-1}$	Given A, to find F $\frac{(1+i)^n-1}{i}$	Given A, to find P $\frac{(1+i)^n-1}{i(1+i)^n}$	n
	5% Compound Interest Factors						6% Compound Interest Factors						
1	1.050	0.9524	1.00000	1.05000	1.000	0.952	1.060	0.9434	1.00000	1.06000	1.000	0.943	1
2	1.103	.9070	0.48780	0.53780	2.050	1.859	1.124	.8900	0.48544	0.54544	2.060	1.833	2
3	1.158	.8638	.31721	.36721	3.153	2.723	1.191	.8396	.31411	.37411	3.184	2.673	3
4	1.216	.8227	.23201	.28201	4.310	3.546	1.262	.7921	.22859	.28859	4.375	3.465	4
5	1.276	.7835	.18097	.23097	5.526	4.329	1.338	.7473	.17740	.23740	5.637	4.212	5
6	1.340	.7462	.14702	.19702	6.802	5.076	1.419	.7050	.14336	.20336	6.975	4.917	6
7	1.407	.7107	.12282	.17282	8.142	5.786	1.504	.6651	.11914	.17914	8.394	5.582	7
8	1.477	.6768	.10472	.15472	9.549	6.463	1.594	.6274	.10104	.16104	9.897	6.210	8
9	1.551	.6446	.09069	.14069	11.027	7.108	1.689	.5919	.08702	.14702	11.491	6.802	9
10	1.629	.6139	.07940	.12950	12.578	7.722	1.791	.5584	.07587	.13587	13.181	7.360	10
11	1.710	.5847	.07039	.12039	14.207	8.306	1.898	.5268	.06679	.12679	14.972	7.887	11
12	1.796	.5568	.06283	.11283	15.917	8.863	2.012	.4970	.05928	.11928	16.870	8.384	12
13	1.886	.5303	.05646	.10646	17.713	9.394	2.133	.4688	.05296	.11296	18.882	8.853	13
14	1.980	.5051	.05102	.10102	19.599	9.899	2.261	.4423	.04758	.10758	21.015	9.295	14
15	2.079	.4810	.04634	.09634	21.579	10.380	2.397	.4173	.04296	.10296	23.276	9.712	15
16	2.183	.4581	.04227	.09227	23.657	10.838	2.540	.3936	.03895	.09895	25.673	10.106	16
17	2.292	.4363	.03870	.08870	25.840	11.274	2.693	.3714	.03544	.09544	28.213	10.477	17
18	2.407	.4155	.03555	.08555	28.132	11.690	2.854	.3503	.03236	.09236	30.906	10.828	18
19	2.527	.3957	.03275	.08275	30.539	12.085	3.026	.3305	.02962	.08962	33.760	11.158	19
20	2.653	.3769	.03024	.08024	33.066	12.462	3.207	.3118	.02718	.08718	36.786	11.470	20
21	2.786	.3589	.02800	.07800	35.719	12.821	3.400	.2942	.02500	.08500	39.993	11.764	21
22	2.925	.3418	.02597	.07597	38.505	13.163	3.604	.2775	.02305	.08305	43.392	12.042	22
23	3.072	.3256	.02414	.07414	41.430	13.489	3.820	.2618	.02128	.08128	46.996	12.303	23
24	3.225	.3101	.02247	.07247	44.502	13.799	4.049	.2470	.01968	.07968	50.816	12.550	24
25	3.386	.2953	.02095	.07095	47.727	14.094	4.292	.2330	.01823	.07823	54.865	12.783	25
26	3.556	.2812	.01956	.06956	51.113	14.375	4.549	.2198	.01690	.07690	59.156	13.003	26
27	3.733	.2678	.01829	.06829	54.669	14.643	4.822	.2074	.01570	.07570	63.706	13.211	27
28	3.920	.2551	.01712	.06712	58.403	14.898	5.112	.1956	.01459	.07459	68.528	13.406	28
29	4.116	.2429	.01605	.06605	62.323	15.141	5.418	.1846	.01358	.07358	73.640	13.591	29
30	4.322	.2314	.01505	.06505	66.439	15.372	5.743	.1741	.01265	.07265	79.058	13.765	30
31	4.538	.2204	.01413	.06413	70.761	15.593	6.088	.1643	.01179	.07179	84.802	13.929	31
32	4.765	.2099	.01328	.06328	75.299	15.803	6.453	.1550	.01100	.07100	90.890	14.084	32
33	5.003	.1999	.01249	.06249	80.064	16.003	6.841	.1462	.01027	.07027	97.343	14.230	33
34	5.253	.1904	.01176	.06176	85.067	16.193	7.251	.1379	.00960	.06960	104.184	14.368	34
35	5.516	.1813	.01107	.06107	90.320	16.374	7.686	.1301	.00897	.06897	111.435	14.498	35
40	7.040	.1420	.00828	.05828	120.800	17.159	10.286	.0972	.00646	.06646	154.762	15.046	40
45	8.985	.1113	.00626	.05626	159.700	17.774	13.765	.0727	.00470	.06470	212.744	15.456	45
50	11.467	.0872	.00478	.05478	209.348	18.256	18.420	.0543	.00344	.06344	290.336	15.762	50
55	14.636	.0683	.00367	.05367	272.713	18.633	24.650	.0406	.00254	.06254	394.172	15.991	55
60	18.679	.0535	.00283	.05283	353.584	18.929	32.988	.0303	.00188	.06188	533.128	16.161	60
65	23.840	.0419	.00219	.05219	456.798	19.161	44.145	.0227	.00139	.06139	719.083	16.289	65
70	30.426	.0329	.00170	.05170	588.529	19.343	59.076	.0169	.00103	.06103	967.932	16.385	70
75	38.833	.0258	.00132	.05132	756.654	19.485	79.057	.0126	.00077	.06077	1,300.949	16.456	75
80	49.561	.0202	.00103	.05103	971.229	19.596	105.796	.0095	.00057	.06057	1,746.600	16.509	80
85	63.254	.0158	.00080	.05080	1,245.087	19.684	141.579	.0071	.00043	.06043	2,342.982	16.549	85
90	80.730	.0124	.00063	.05063	1,594.607	19.752	189.465	.0053	.00032	.06032	3,141.075	16.579	90
95	103.035	.0097	.00049		2,040.694	19.806	253.546	.0039	.00024	.06024	4,209.104	16.601	95
100	131.501	.0076	.00038	.05038	2,610.025	19.848	339.302	.0029	.00018	.06018	5,638.368	16.618	100

*Factors presented for two interest rates only. By using the appropriate formulas, values for other interest rates may be calculated.

TABLE 9-22 Examples of the Use of Compound Interest Table

Given: \$2500 is invested now at 5 percent.
Required: Accumulated value in 10 years (i.e., the amount of a given principal).

Solution:

$$F = P(1+i)^n = \$2500 \times 1.05^{10}$$
$$\text{Compound-amount factor} = (1+i)^n = 1.05^{10} = 1.629$$
$$F = \$2500 \times 1.629 = \$4062.50$$

Given: \$19,500 will be required in 5 years to replace equipment now in use.
Required: With interest available at 3 percent, what sum must be deposited in the bank at present to provide the required capital (i.e., the principal which will amount to a given sum)?

Solution:

$$P = F\frac{1}{(1+i)^n} = \$19{,}500\frac{1}{1.03^5}$$
$$\text{Present-worth factor} = 1/(1+i)^n = 1/1.03^5 = 0.8626$$
$$P = \$19{,}500 \times 0.8626 = \$16{,}821$$

Given: \$50,000 will be required in 10 years to purchase equipment.
Required: With interest available at 4 percent, what sum must be deposited each year to provide the required capital (i.e., the annuity which will amount to a given fund)?

Solution:

$$A = F\frac{i}{(1+i)^n - 1} = \$50{,}000\frac{0.04}{1.04^{10} - 1}$$
$$\text{Sinking-fund factor} = \frac{i}{(1+i)^n - 1} = \frac{0.04}{1.04^{10} - 1} = 0.08329$$
$$A = \$50{,}000 \times 0.08329 = \$4{,}164$$

Given: \$20,000 is invested at 10 percent interest.
Required: Annual sum that can be withdrawn over a 20-year period (i.e., the annuity provided by a given capital).

Solution:

$$A = P\frac{i(1+i)^n}{(1+i)^n - 1} = \$20{,}000\frac{0.10 \times 1.10^{20}}{1.10^{20} - 1}$$
$$\text{Capital-recovery factor} = \frac{i(1+i)^n}{(1+i)^n - 1} = \frac{0.10 \times 1.10^{20}}{1.10^{20} - 1} = 0.11746$$
$$A = \$20{,}000 \times 0.11746 = \$2349.20$$

Given: \$500 is invested each year at 8 percent interest.
Required: Accumulated value in 15 years (i.e., amount of an annuity).

Solution:

$$F = A\frac{(1+i)^n - 1}{i} = \$500\frac{1.08^{15} - 1}{0.08}$$
$$\text{Compound-amount factor} = \frac{(1-i)^n - 1}{i} = \frac{1.08^{15} - 1}{0.08} = 27.152$$
$$F = \$500 \times 27.152 = \$13{,}576$$

Given: \$8000 is required annually for 25 years.
Required: Sum that must be deposited now at 6 percent interest.

Solution:

$$P = A\frac{(1+i)^n - 1}{i(1+i)^n} = \$8000\frac{1.06^{25} - 1}{0.06 \times 1.06^{25}}$$
$$\text{Present-worth factor} = \frac{(1+i)^n - 1}{i(1+i)^n} = \frac{1.06^{25} - 1}{0.06 \times 1.06^{25}} = 12.783$$
$$P = \$8000 \times 12.783 = \$102{,}264$$

sketch of this process. The time periods are years, and the interest is normally on an annual basis using end-of-year money flows. The longer the time before money is received, the less it is worth at present.

Effective Interest Rates When an interest rate is quoted, it is *nominal* interest that is stated. These quotes are on an annual basis, however, when compounding occurs that is not the actual or effective interest. According to government regulations, an effective rate APY must be stated also. The effective interest is calculated by

$$i_{\text{eff}} = \left(1 + \frac{i}{m}\right)^{(m)(1)} - 1 \qquad (9\ 16)$$

The time period for calculating the effective interest rate is 1 year.

TABLE 9-23 Condensed Continuous Interest Table*

Factors for determining zero-time values for cash flows which occur at other than zero time.

Compounding of Cash Flows Which Occur:	1%	5%	10%	15%	20%	25%	30%	35%	40%	50%	60%	70%	80%	90%	100%
A. In an Instant															
½ year before	1.005	1.025	1.051	1.078	1.105	1.133	1.162	1.191	1.221	1.284	1.350	1.419	1.492	1.568	1.649
1 " "	1.010	1.051	1.105	1.162	1.221	1.284	1.350	1.419	1.492	1.649	1.822	2.014	2.226	2.460	2.718
1½ " "	1.015	1.078	1.162	1.252	1.350	1.455	1.568	1.690	1.822	2.117	2.460	2.858	3.320	3.857	4.482
2 " "	1.020	1.105	1.221	1.350	1.492	1.649	1.822	2.014	2.226	2.718	3.320	4.055	4.953	6.050	7.389
3 " "	1.030	1.162	1.350	1.568	1.822	2.117	2.460	2.858	3.320	4.482	6.050	8.166	11.023	14.880	20.086
B. Uniformly until Zero Time															
From ½ year before to 0 time	1.002	1.013	1.025	1.038	1.052	1.065	1.079	1.093	1.107	1.136	1.166	1.197	1.230	1.263	1.297
" 1 " " " " "	1.005	1.025	1.052	1.079	1.107	1.136	1.166	1.197	1.230	1.297	1.370	1.448	1.532	1.622	1.718
" 1½ " " " " "	1.008	1.038	1.079	1.121	1.166	1.213	1.263	1.315	1.370	1.489	1.622	1.769	1.933	2.117	2.321
" 2 " " " " "	1.010	1.052	1.107	1.166	1.230	1.297	1.370	1.448	1.532	1.718	1.933	2.182	2.471	2.805	3.194
" 3 " " " " "	1.015	1.079	1.166	1.263	1.370	1.489	1.622	1.769	1.933	2.321	2.805	3.412	4.176	5.141	6.362
Discounting of Cash Flows Which Occur:															
C. In an Instant															
1 year later	.990	.951	.905	.861	.819	.779	.741	.705	.670	.606	.549	.497	.449	.407	.360
2 " "	.980	.905	.819	.741	.670	.606	.549	.497	.449	.368	.301	.247	.202	.165	.135
3 " "	.970	.861	.741	.638	.549	.472	.407	.350	.301	.223	.165	.122	.091	.067	.050
4 " "	.961	.819	.670	.549	.449	.368	.301	.247	.202	.135	.091	.061	.041	.027	.018
5 " "	.951	.779	.606	.472	.368	.286	.223	.174	.135	.082	.050	.030	.018	.011	.007
10 years later	.905	.606	.368	.223	.135	.082	.050	.030	.018	.007	.002	.001	—	—	—
15 " "	.861	.472	.223	.105	.050	.024	.011	.005	.002	.001	—	—	—	—	—
20 " "	.819	.368	.135	.050	.018	.007	.002	.001	—	—	—	—	—	—	—
25 " "	.779	.286	.082	.024	.007	.002	.001	—	—	—	—	—	—	—	—
D. Uniformly over Individual Years															
1st year	.995	.975	.952	.929	.906	.885	.864	.844	.824	.787	.752	.719	.688	.659	.632
2nd "	.985	.928	.861	.799	.742	.689	.640	.595	.552	.477	.413	.357	.309	.268	.232
3rd "	.975	.883	.779	.688	.608	.537	.474	.419	.370	.290	.226	.177	.139	.109	.086
4th "	.966	.840	.705	.592	.497	.418	.351	.295	.248	.176	.124	.088	.062	.044	.032
5th "	.956	.799	.638	.510	.407	.326	.260	.208	.166	.106	.068	.044	.028	.018	.012
6th year	.946	.760	.577	.439	.333	.254	.193	.147	.112	.065	.037	.022	.013	.007	.004
7th "	.937	.723	.522	.378	.273	.197	.143	.103	.075	.039	.020	.011	.006	.003	.002
8th "	.928	.687	.473	.325	.224	.154	.108	.073	.050	.024	.011	.005	.002	.001	.001
9th "	.918	.654	.428	.280	.183	.120	.078	.051	.034	.014	.006	.003	.001	—	—
10th "	.909	.622	.387	.241	.150	.093	.058	.036	.022	.009	.003	.001	—	—	—
E. Uniformly over 5-Year Periods															
1st 5 years	.975	.885	.787	.704	.632	.571	.518	.472	.432	.367	.317	.277	.245	.220	.199
6th through 10th year	.928	.689	.477	.332	.232	.164	.116	.082	.058	.030	.016	.008	.004	.002	.001
11th through 15th year	.883	.537	.290	.157	.086	.047	.026	.014	.008	.002	.001	—	—	—	—
16th through 20th year	.840	.418	.176	.074	.032	.013	.006	.002	.001	—	—	—	—	—	—
21st through 25th year	.799	.326	.106	.035	.012	.004	.001	—	—	—	—	—	—	—	—
F. Declining to Nothing at Constant Rate															
1st 5 years	.983	.922	.852	.791	.736	.687	.643	.603	.568	.506	.456	.413	.377	.347	.320
" 10 "	.968	.852	.736	.643	.568	.506	.456	.413	.377	.320	.278	.245	.219	.198	.180
" 15 "	.952	.791	.643	.536	.456	.394	.347	.309	.278	.231	.198	.172	.153	.137	.124
" 20 "	.936	.736	.568	.456	.377	.320	.278	.245	.219	.180	.153	.133	.117	.105	.095
" 25 "	.922	.687	.506	.394	.320	.269	.231	.203	.180	.147	.124	.108	.095	.085	.077

*From tables compiled by J. C. Gregory, The Atlantic Refining Co.

TABLE 9-24 Summary of Discrete and Compound Interest Equations

Factor	Find	Given	Discrete compounding		Continuous compounding	
Single payment						
Compound amount	F	P	$F = P(1+i)^n$	$P(F/P\ i,n)$	$F = P(e^{rn})$	$P(F/P\ r,n)^\infty$
Present worth	P	F	$P = F\left[\frac{1}{(1+i)^n}\right]$	$F(P/F\ i,n)$	$P = F(e^{-rn})$	$F(P/F\ r,n)^\infty$
Uniform series						
Compound amount	F	A	$F = A\left[\frac{(1+i)^n - 1}{i}\right]$	$A(F/A\ i,n)$	$F = A\left(\frac{e^{rn}-1}{e^r - 1}\right)$	$F(F/A\ r,n)^\infty$
Sinking fund	A	F	$A = F\left[\frac{i}{(1+i)^n - 1}\right]$	$F(A/F\ i,n)$	$A = F\left(\frac{e^r - 1}{e^{rn} - 1}\right)$	$F(A/F\ r,n)^\infty$
Present worth	P	A	$P = A\left[\frac{(1+i)^n - 1}{i(1+i)^n}\right]$	$A(F/A\ i,n)$	$P = A\left[\frac{e^{rn}-1}{e^{rn}(e^r - 1)}\right]$	$A(P/A\ r,n)^\infty$
Capital recovery	A	P	$A = P\left[\frac{i(1+i)^n}{(1+i)^n - 1}\right]$	$P(A/P\ i,n)$	$A = P\left[\frac{e^{rn}(e^r - 1)}{e^{rn} - 1}\right]$	$A(P/A\ r,n)^\infty$

Example 6: Effective Interest Rate A person is quoted an 8.33 percent nominal interest rate on a 4-year loan compounded monthly. Determine the effective interest rate.

Solution:

$$i_{\text{eff}} = \left(\frac{1+i}{m}\right)^{(m)(1)} = \left(\frac{1+0.833}{12}\right)^{(12)(1)} - 1 = (1.00694)^{12} - 1$$

$$= 1.0865 - 1 = 0.0865$$

The effective interest rate is 8.65 percent.

CASH FLOW

Cash flow is the amount of funds available to a company to meet current operating expenses. Cash flow may be expressed on a before- or after-tax basis. *After-tax cash flow* is defined as the net profit (income) after taxes plus depreciation. It is an integral part of the net present worth (NPW) and discounted cash flow profitability calculations.

The cash flow diagram, also referred to as a cash flow model (Fig. 9-9), shows the relationship between revenue, cash operating expenses, depreciation, and profit. This diagram is similar in many respects to a process flow diagram, but it is in dollars. Revenue is generated from the sale of a product manufactured in "operations." Working capital is replenished from sales and may be considered to be in dynamic equilibrium with operations. Leaving the operations box is a stream, "cash operating expenses." It includes all the cash expenses incurred in the operation but does not include the noncash item depreciation. Since depreciation is an allowance, it is reported on the operating expense sheet, in accordance with the tax laws, as an operating expense item. (See the section "Operating Expense Estimation.") Depreciation is an internal expense, and this allowance is retained within the company. If the cash operating expenses are subtracted from the revenue, the result is the operating income. If depreciation is subtracted from the operating income, the net profit before taxes results. Federal income taxes are then deducted from the net profit before taxes, giving the net profit after taxes. When depreciation and net profit after taxes are summed, the result is the after-tax cash flow. The terminology in Fig. 9-9 is consistent with that found in most company income statements in company annual reports.

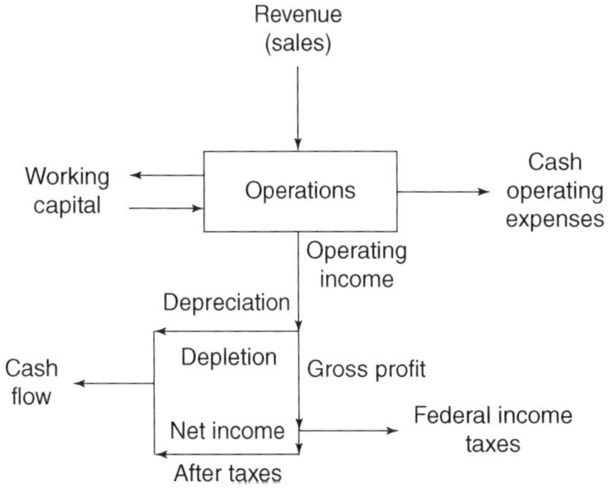

FIG. 9-9 Cash flow model.

An equation can be developed for cash flow as follows:

$$CF = (R - C - D)(1 - t) + D \tag{9-17}$$

where R = revenue
C = cash operating expenses
D = depreciation
t = tax rate
CF = after-tax cash flow

Equation (9-17) can be rearranged algebraically to yield Eq. (9-18)

$$CF = t(D) + (1 - t)(\Delta R) - (1 - t)(\Delta C) \tag{9-18}$$

The term $t \times D$ is only the result of an algebraic manipulation, and no interpretation should be assumed. This term $t \times D$ is the contribution to cash flow from depreciation, and $(1 - t) \times R$ and $(1 - t) \times C$ are the contributions to cash flow from revenues and cash operating expenses, respectively. Example 7 is a sample calculation of the after-tax cash flow and the tabulated results.

Example 7: After-Tax Cash Flow The revenue from the manufacture of a product in the first year of operation is \$9.0 million, and the cash operating expenses are \$4.5 million. Depreciation on the invested capital is \$1.7 million. If the federal income tax rate is 35 percent, calculate the after-tax cash flow.

Solution: The resulting after-tax cash flow is \$3.52 million. See Fig. 9-10.

Cumulative Cash Position Table To organize cash flow calculations, it is suggested that a cumulative cash position table be prepared by using an electronic spreadsheet. For this discussion,

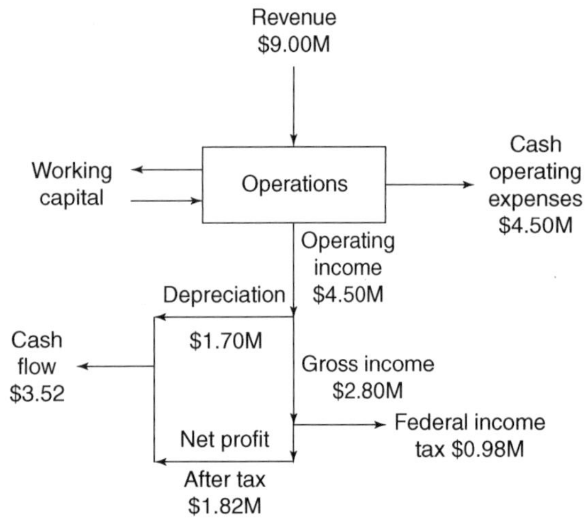

FIG. 9-10 Cash flow model for Example 7. M = million.

time zero is assumed to be at project start-up. Expenditures for land and equipment occurred prior to time zero and represent negative cash flows. At time zero, working capital is charged to the project as a negative cash flow. Start-up expenses are charged in the first year, and positive cash flow from the sale of product as net income after taxes plus depreciation begins, reducing the negative cash position. This process continues until the project is terminated. At that time, adjustments are made to recover land and working capital. An example of a cumulative cash position table is Table 9-25.

When equipment is added for plant expansions to an existing facility, it may be more convenient to use time zero when the first expenditures occur. The selection of either time base is satisfactory for economic analysis as long as consistency is maintained.

Example 8: Cumulative Cash Position Table (Time Zero at Start-up) A specialty chemical company is considering the manufacture of an additive for use in the plastics industry. The following is a list of production, sales, and cash operating expenses.

Year	Production, Mlb	Sales, $1,000	Cash operating expenses, $1,000
1	40	20,000	10,320
2	42	21,000	10,800
3	45	23,400	11,520
4	48	24,960	12,240
5	50	27,500	13,470
6	50	28,000	13,970
7	47	23,500	13,175
8	45	21,600	12,645
9	40	18,800	11,320
10	35	15,750	9,995

Land for the project is available at $300,000. The fixed capital investment was estimated to be $12,000,000. A working capital of $1,800,000 is needed initially for the venture. Start-up expenses based upon past experience are estimated to be $750,000. The project qualifies under IRS guidelines as a 5-year class life investment. The company uses MACRS depreciation with the half-year convention. At the conclusion of the project, the land and working capital are returned to management. Develop a cash flow analysis for this project, using a cumulative cash position table (Table 9-25).

Cumulative Cash Position Plot A pictorial representation of the cumulative cash flows as a function of time is the cumulative cash position plot. All expenditures for capital as well as revenue from sales are plotted as a function of time. Figure 9-11 is such an idealized plot showing time zero at start-up in part *a* and time zero when the first funds are expended in part *b*. It should be understood that the plots have been idealized for illustration purposes. Expenditures are usually stepwise, and accumulated cash flow from sales is seldom a straightline but more likely a curve with respect to time.

Time Zero at Start-up Prior to time zero, expenditures are made for land, fixed capital investment, and working capital. It is assumed that land had been purchased by the company at some time in the past, and a parcel is allocated for the project under consideration. Land is allocated instantaneously to the project sometime prior to the purchase of equipment and construction of the plant. The fixed capital investment is purchased and installed over a period of time prior to start-up. For the purpose of this presentation, it is assumed that it occurs uniformly over a period of time. Both land and fixed capital investment are compounded to time zero by using the appropriate compound interest factors. At time zero, working capital is charged to the project. Start-up expenses are entered in the first year of operation after start-up. After time zero, start-up occurs and then manufacturing begins and income is generated, so cash flow begins to accumulate if the process is sound. At the end of the project life, land and working capital are recovered instantaneously.

TABLE 9-25 Cash Flow Analysis for Example 8

Year	−2	−2 to 0	0	1	2	3	4	5	6	7	8	9	10	End 10
Production Mlb/yr				40	42	45	48	50	50	47	45	40	35	
All money is $1,000														
Land	−300													
Fixed capital investment $1000		−12,000												
Working capital			−1,800											
Start-up expenses			−750											
Total capital investment			14,250											
Sales				20,000	21,000	23,400	24,960	27,500	28,000	23,500	21,600	18,800	15,750	
Cash operating expenses				10,320	10,800	11,520	12,240	13,470	13,970	13,175	12,645	11,320	9,995	
Operating income				9,680	10,200	11,880	12,720	14,030	14,030	10,325	8,955	7,480	5,755	
Depreciation				2,400	3,840	2,300	1,390	1,380	690	0	0	0	0	
Net income before taxes				7,280	6,360	9,580	11,330	12,650	13,340	10,325	8,955	7,480	5,755	
Income tax				3,640	2,226	3,353	3,966	4,428	4,669	3,614	3,134	2,618	2,014	
Net income after taxes				3,640	4,134	6,227	7,364	8,222	8,671	6,711	5,821	4,862	3,741	
Depreciation				2,400	3,840	2,300	1,390	1,380	690	0	0	0	0	
After-tax cash flow	−300	−12,000	−2,550	6,040	7,974	8,527	8,754	9,602	9,361	6,711	5,821	4,882	3,741	
Cumulative cash flow	−300	−12,300	−14,850	−8,810	−836	7,691	16,445	26,047	35,408	42,119	47,940	52,802	56,543	
Capital recovery														2,100
End of project value														58,643

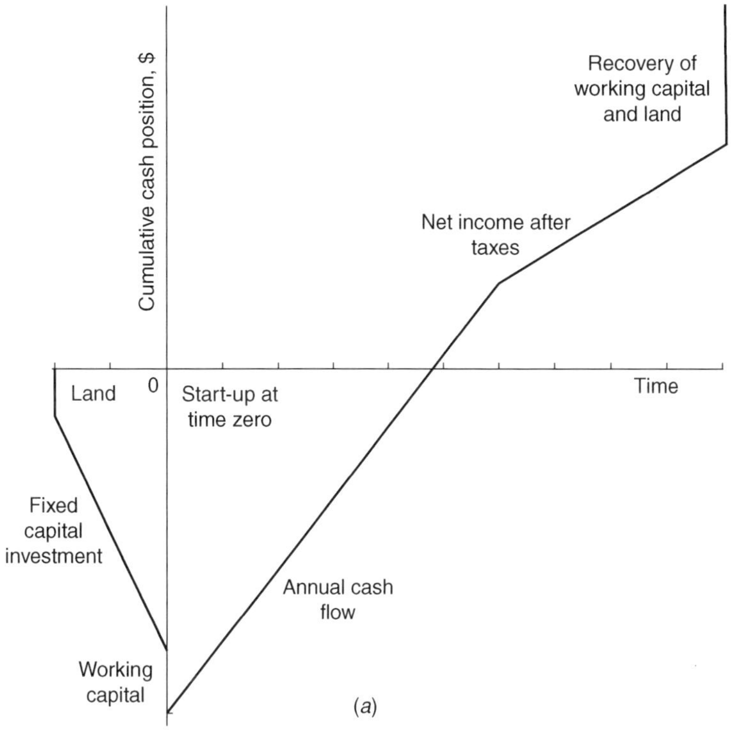

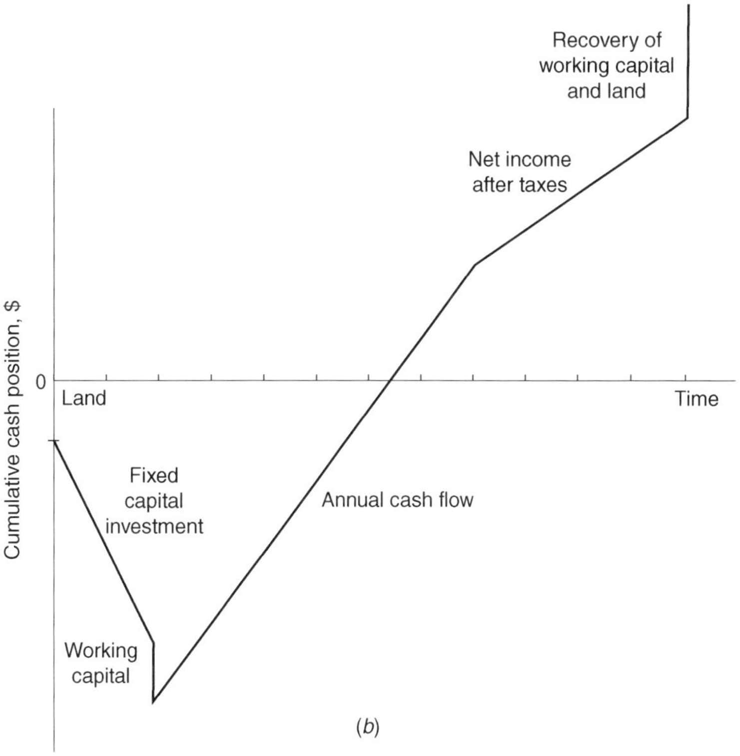

FIG. 9-11 Typical cumulative cash position plot. (*a*) Time zero is start-up. (*b*) Time zero occurs when first funds are spent.

PROFITABILITY

In the free enterprise system, companies are in business to make a profit. Management has the responsibility of investing in ventures that are financially attractive by increasing earnings, providing attractive rates of return, and increasing value added. Every viable business has limitations on the capital available for investment; therefore, it will invest in the most economically attractive ventures. The objectives and goals of a company are developed by management. Corporate objectives may include one or several of the following: maximize return on investment, maximize return on stockholders' equity, maximize aggregate earnings, maximize common stock prices, increase market share, increase economic value, increase earnings per share of stock, and increase market value added. These objectives are the ones most frequently listed by executives.

To determine the worthiness of a venture, quantitative and qualitative measures of profitability are considered.

QUANTITATIVE MEASURES OF PROFITABILITY

When a company invests in a venture, the investment must earn more than the cost of capital for it to be worthwhile. A profitability estimate is an attempt to quantify the desirability of taking a risk in a venture.

The *minimum acceptable rate of return (MARR)* for a venture depends on a number of factors such as interest rate, cost of capital, availability of capital, degree of risk, economic project life, and other competing projects. Management will adjust the MARR depending on any of the above factors to screen out the more attractive ventures. When a company invests in a venture, the investment must earn more than the cost of capital and should be able to pay dividends.

Although there have been many quantitative measures suggested through the years, some did not take into account the time value of money. In today's economy, the following measures are the ones most companies use:

Payout period (POP) plus interest
Net present worth (NPW)
Discounted cash flow (DCFROR)

Payout Period Plus Interest Payout period (POP) is the time that will be required to recover the depreciable fixed capital investment from the accrued after-tax cash flow of a project with no interest considerations. In equation format

$$\text{Payout period} = \frac{\text{depreciable fixed capital investment}}{\text{after-tax cash flow}} \tag{9-19}$$

This model does not take into account the time value of money, and no consideration is given to cash flows that occur in a project's later years after the depreciable investment has been recovered. A variation on this method includes interest, called payout period plus interest (POP + I); and the net effect is to increase the payout period. This variation accounts for the time value of money.

$$\text{Payout period plus interest (POP + I)} = \left(\frac{\text{depreciable fixed capital investment}}{\text{after-tax cash flow}}\right)_i \tag{9-20}$$

Neither of these methods makes provision for including land and working capital, and no consideration is given to cash flows that occur in a project's later years after the depreciable fixed investment has been recovered for projects that earn most of their profit in the early years.

Net Present Worth In the net present worth method, an arbitrary time frame is selected as the basis of calculation. This method is the measure many companies use, as it reflects properly the time value of money and its effect on profitability. In equation form

$$\text{Net present worth (NPW)} = \text{present worth of all cash inflows} - \text{present worth of all investment items} \tag{9-21}$$

When the NPW is calculated according to Eq. (9-21), if the result is positive, the venture will earn more than the interest (discount) rate used; conversely, if the NPW is negative, the venture earns less than that rate.

Discounted Cash Flow In the discounted cash flow method, all the yearly after-tax cash flows are discounted or compounded to time zero depending upon the choice of time zero. The following equation is used to solve for the interest rate *i*, which is the discounted cash flow rate of return (DCFROR).

$$\text{DCFROR} = \sum_0^n \text{(after-tax cash flows)} = 0 \tag{9-22}$$

Equation (9-22) may be solved graphically or analytically by an iterative trial-and-error procedure for the value of *i*, which is the discounted cash flow rate of return. It has also been known as the *profitability index*. For a project to be profitable, the interest rate must exceed the cost of capital.

The effect of interest on the cash position of a project is shown in Fig. 9-12. As interest increases, the time to recover the capital expenditures is increased.

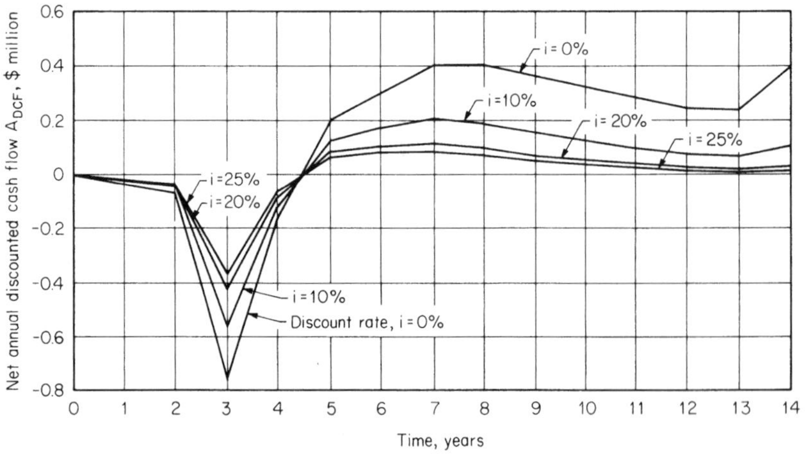

FIG. 9-12 Effect of interest rate on cash flow (time zero occurs when first funds are expanded).

TABLE 9-26 Profitability Analysis for Example 9

Time	Cash flow	20% interest factors	Present worth 20% interest	30% interest factors	Present worth 30% interest	40% interest factors	Present worth 40% interest
−2	−300	1.492	−448	1.822	−547	2.226	−668
−2 to 0	−12,000	1.230	−14,760	1.370	−16,440	1.532	−18,384
0	−2,250	1.000	−2,250	1.000	−2,250	1.000	−2,250
1	6,040	0.906	5,472	0.864	5,219	0.824	4,977
2	7,974	0.742	5,917	0.640	5,103	0.552	4,402
3	8,527	0.608	5,184	0.474	4,042	0.370	3,155
4	8,754	0.497	4,351	0.351	3,073	0.248	2,171
5	9,602	0.407	3,908	0.260	2,497	0.166	1,594
6	9,361	0.333	3,117	0.193	1,807	0.112	1,048
7	6,711	0.273	1,832	0.143	960	0.075	503
8	5,821	0.224	1,304	0.106	617	0.050	291
9	4,862	0.183	890	0.078	379	0.034	165
10	3,471	0.150	561	0.058	217	0.022	82
End 10	2,100	0.135	284	0.050	105	0.018	284
Net present worth			15,362		4,782		−2,360
Discounted cash flow rate of return		33.90%					

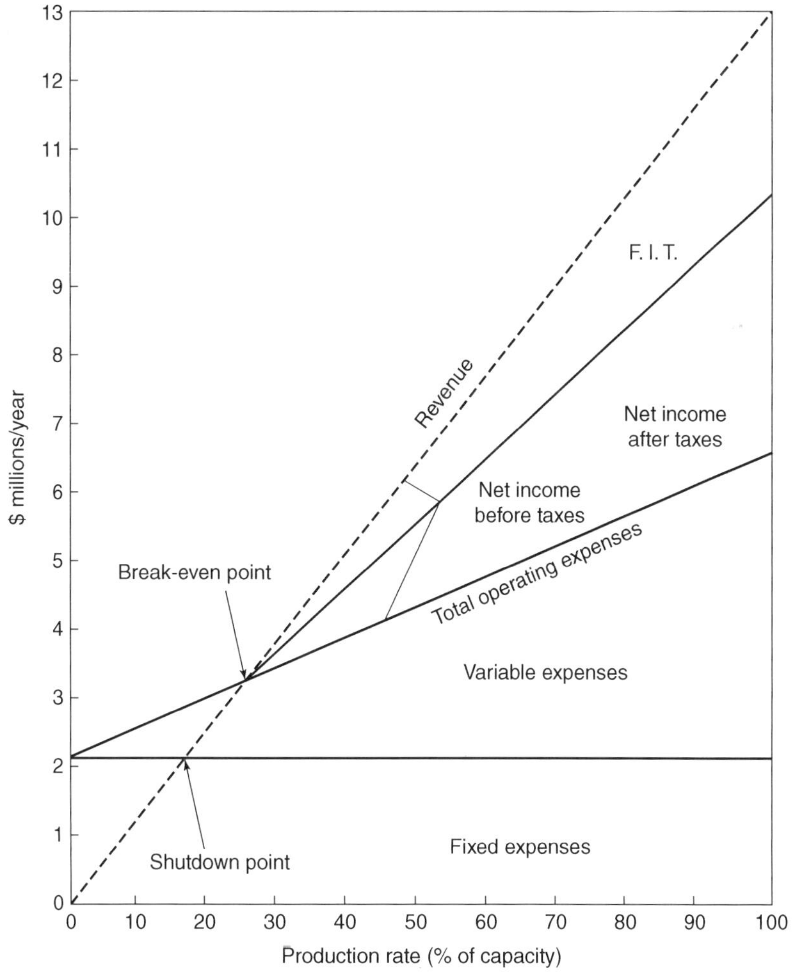

FIG. 9-13 Break-even plot.

In the chemical business, operating net profit and cash flow are received on a nearly continuous basis, therefore, there is justification for using the condensed continuous interest tables, such as Table 9-23, in discounted cash flow calculations.

Example 9: Profitability Calculations Example 8 data are used to demonstrate these calculations. Calculate the following:

a. Payout period (POP)
b. Payout period with interest (POP + I)
c. NPW at a 30 percent interest rate
d. DCF rate of return

Solution:

a. From Table 9-26, the second column is the cash flow by years with no interest. The payout period occurs where the cumulative cash flow is equal to the fixed capital investment, $12,000,000 or 1.7 years.

b. In Table 9-26, the payout period at 30 percent interest occurs at 2.4 years.

c. The results of the present worth calculations for 20, 30, and 40 percent interest rates are tabulated. At 30 percent interest, the net present worth is $4,782,000, and since it is a positive figure, this means the project will earn more than 30 percent interest.

d. Discounted cash flow rate of return is determined by interpolating in Table 9-26. At 30 percent interest the net present worth is positive, and at 40 percent interest it is negative. By definition, the DCFROR occurs when the summation of the net present worth equals zero. This occurs at an interest of 33.9 percent.

QUALITATIVE MEASURES

In addition to quantitative measures, there are certain qualitative measures or intangible factors that may affect the ultimate investment decision. Those most frequently mentioned by management are employee morale, employee safety, environmental constraints, legal constraints, product liability, corporate image, and management goals. Attempts have been made to quantify these intangibles by using an ordinal or a ranking system, but most have had little or no success. Couper (2003) discussed in greater detail the effect of qualitative measures on the decision-making process.

SENSITIVITY ANALYSIS

Whenever an economic study is prepared, the marketing, capital investment, and operating expense data used are estimates, and therefore a degree of uncertainty exists. Questions arise such as, What if the capital investment is 15 percent greater than the value reported? A sensitivity analysis is used to determine the effect of percentage changes in pertinent variables on the profitability of the project. Such an analysis indicates which variables are most susceptible to change and need further study.

Break-Even Analysis Break-even analysis is a simple form of sensitivity analysis and is a useful concept that can be of value to managers. *Break-even* refers to the point in an operation where income just equals expenses. Figure 9-13 is a pictorial example of the results of a break-even analysis, showing that the break-even point is at 26 percent of production capacity. Management wants to do better than just break even; therefore, such plots can be used as a profit planning tool, for product pricing, production operating level, incremental equipment costs, etc. Another significant point is the *shutdown point* where revenue just equals the fixed expenses. Therefore, if a proposed operation can't make fixed expenses, it should be shut down.

Strauss Plot R. Strauss (*Chem. Eng.*, pp. 112–116, Mar. 25, 1968) developed a sensitivity plot, in Fig. 9-14, in which the ordinate is a measure of profitability and the abscissa is the change in a variable greater than (or less than) the value used in the base case. Where the abscissa crosses the ordinate is the result of the base case of NPW, return, annual worth, etc. The slope of a line on this "spider" plot is the degree of change in profitability resulting from a change in a

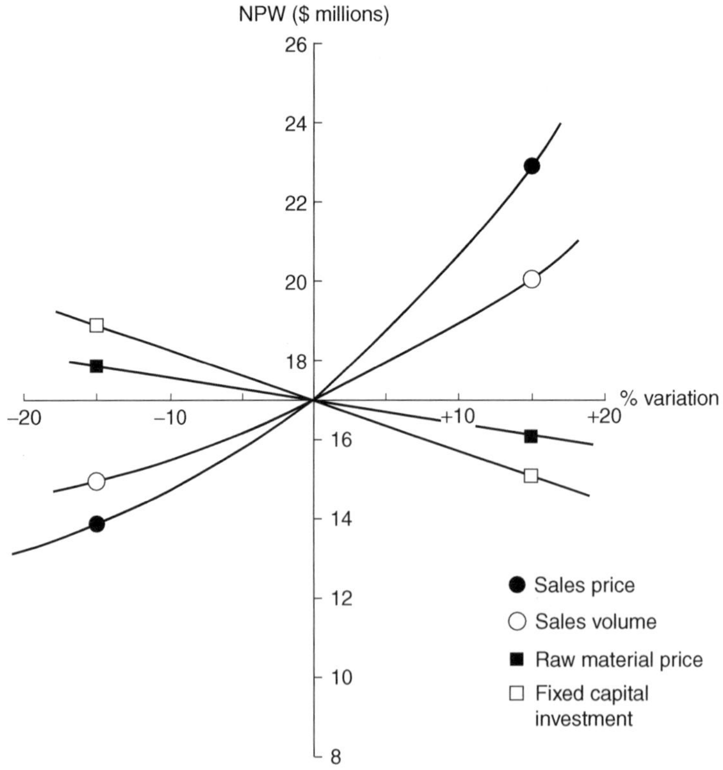

FIG. 9-14 Strauss plot.

variable, selling price, sales volume, investment, etc. The length of the line represents the sensitivity of the variable and its degree of uncertainty. Positive-slope lines are income-related, and negative-slope lines are expense-related. A spreadsheet is useful in developing data for this "what if" plot since numerous scenarios must be prepared to develop the plot.

Tornado Plot Another graphical sensitivity analysis is the "tornado" plot. Its name is derived from the shape of the resulting envelope. As in other methods, a base case is solved first, usually expressing the profitability as the net present worth. In Fig. 9-15, the NPW is a vertical line, and variations in each selected variable above and below the base case are solved and plotted. In this figure, the variables of selling price, sales volume, operating expenses, raw material expenses, share of the market, and investment are plotted. It is apparent that the selling price and sales volume are the critical factors affecting the profitability. A commercial computer program known as @RISK® developed by the Palisade Corporation, Newfield, N.Y., may be used to prepare a tornado plot.

Relative Sensitivity Plot Another type of analysis developed by J. C. Agarwal and I. V. Klumpar (*Chem. Eng.*, pp. 66–72, Sept. 29, 1975) is the relative sensitivity plot. The variables studied are related to those in the base case, and the resulting plot is the relative profitability.

Although sensitivity analyses are easy to prepare and they yield useful information for management, there is a serious disadvantage. Only one variable at a time can be studied. Frequently, there are synergistic effects among variables; e.g., in marketing, the variables such as sales volume, selling price, and market share may have a synergistic effect, and that effect cannot be taken into account. Other interrelated variables such as fixed capital investment, maintenance, and other investment-based items also cannot be represented properly. These disadvantages lead to another management tool—uncertainty analysis.

UNCERTAINTY ANALYSIS

This analysis allows the user to account for variable interaction that is another level of sophistication. Two terms need clarification—

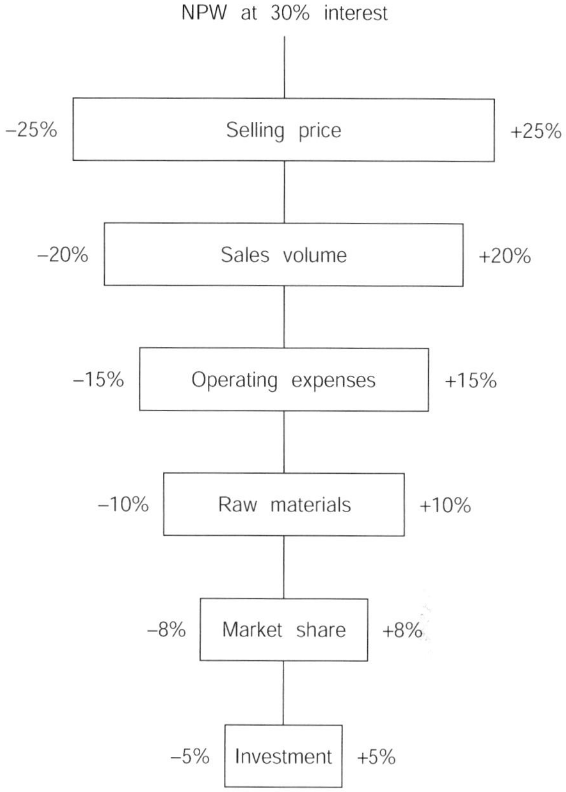

FIG. 9-15 Typical tornado plot. (*Source: Adapted from Couper, 2003.*)

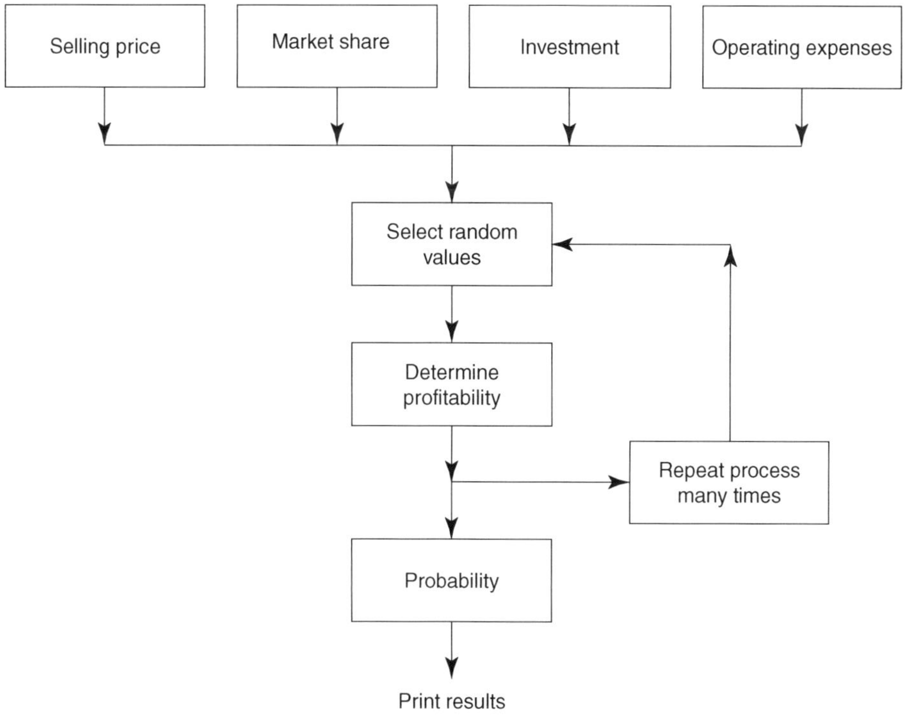

FIG. 9-16 Schematic diagram of Monte Carlo simulation.

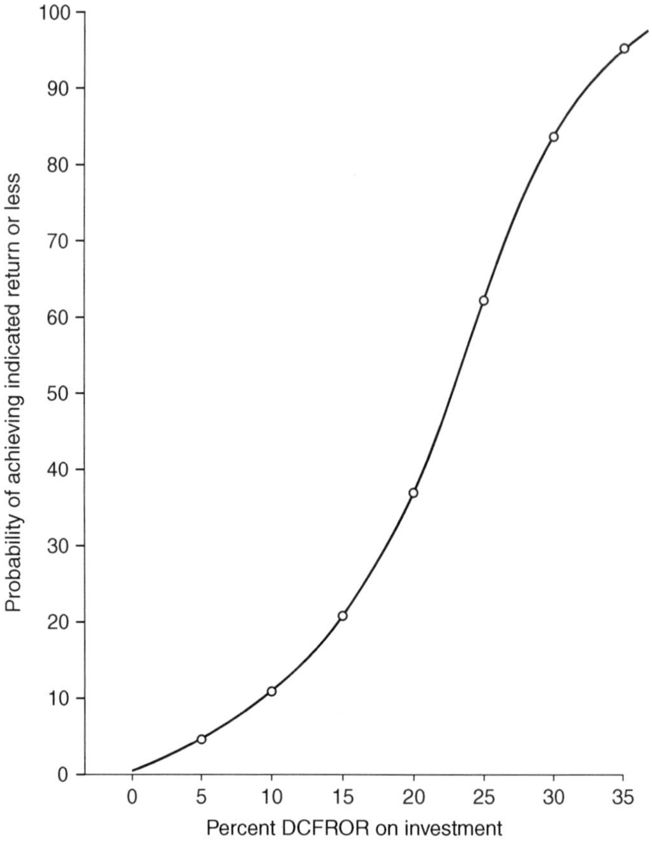

FIG. 9-17 Probability curve for Monte Carlo simulation.

uncertainty and *risk*. Uncertainty is exactly what the word means—not certain. Risk, however, implies that the probability of achieving a specific outcome is known within certain confidence limits.

Since sensitivity analysis has the shortcoming of being able to inspect only one variable at a time, the next step is to use probability risk analysis, generally referred to as the Monte Carlo technique. R. C. Ross (*Chem. Eng.*, pp. 149–155, Sept. 20, 1971), P. Macalusa (*BYTE*, pp. 179–192, March 1984), and D. B. Hertz (*Harvard Bus. Rev.*, pp. 96–108, Jan-Feb 1968) have written classic articles on the use of the Monte Carlo technique in uncertainty analysis. These articles incorporate subjective probabilities and assumptions of the distribution of errors into the analysis. Each variable is represented by a probability distribution model. Figure 9-16 is a pictorial representation of the steps in the Monte Carlo simulation. The first step is to gather enough data to develop a reasonable probability model. Not all variables follow the normal distribution curve, but perhaps sales volume and sales-related variables do. Studies have shown that capital investment estimates are best represented by a beta distribution. Next the task is to select random values from the various models by using a random number generator and from these data calculate a profitability measure such as NPW or rate of return. The procedure is repeated a number of times to generate a plot of the probability of achieving a given profitability versus profitability. Figure 9-17 is a typical plot. Once the analysis has been performed, the next task is to interpret the results. Management must understand what the results mean and the reliability of the results. Experience can be gained only by performing uncertainty analyses, not just one or two attempts, to develop confidence in the process. The stakes may be high enough to spend time and learn the method. Software companies such as @RISK or SAS permit the user to develop probability models and perform the Monte Carlo analysis. The results may be plotted as the probability of achieving at least a given return or of achieving less than the desired profitability.

FEASIBILITY ANALYSIS

A feasibility analysis is prepared for the purpose of determining that a proposed investment meets the minimum requirements established

TABLE 9-27 Checklist of Required Information for a Feasibility Analysis

Fixed capital investment
Working capital requirements
Total capital investment
Total manufacturing expense
Packaging and in-plant expense
Total product expense
General overhead expense
Total operating expense
Marketing data
Cash flow analysis
Project profitability
Sensitivity analysis
Uncertainty analysis

TABLE 9-28 Marketing Data Template

Project title:
Basis: Sales and marketing data are not inflated (20__ dollars)

	20__		20__		20__	
	Amount	%Total	Amount	%Total	Amount	%Total
Total market						
Units						
Average realistic price, $/unit						
Value, $						
Estimated product sales (with AR)						
Units						
Average realistic price, $/unit						
Value, $						
Current product sales (without AR)						
Units						
Average realistic price, $/unit						
Value, $						
Incremental product sales						
Units						
Average realistic price, $/unit						
Value, $						
Current product sales displaced by improved product sales						
Units						
Value, $						
Total improved product sales						
Units						
Value, $						

NOTE: Table extends to the right to accommodate the number of project years.
AR = appropriation request.

TABLE 9-29 Cash Flow Analysis Template

	Cash flow summary		
	200X	200Y	200Z, etc.*
Investment			
Land			
Fixed capital investment			
Offsite capital			
Allocated capital			
Working capital			
Start-up expenses			
Interest			
Catalysts and chemicals			
Licenses, patents, etc.			
Total capital investment			
Income statement			
Income			
Expenses			
Cash operating expenses			
Depreciation			
Total operating expenses			
Operating income			
Net income before taxes			
Federal income taxes			
Net income after taxes			
Cash flow			
Capital recovery			
Cumulative cash flow			

*Table may be extended to the right to accommodate the number of years of the project.

by management. It should be in sufficient detail to provide management with the facts required to make an investment decision. All the basic information has been discussed in considerable detail in the earlier parts of Sec. 9.

The minimum information required should include, but not be limited to, that in Table 9-27. Forms and spreadsheets are the most succinct method to present the information. The forms should state clearly the fund amounts and the date that each estimate was performed. The forms may be developed so that data for other scenarios may be reported by extending the tables to the right of the page. It is suggested that blank lines be included for any additional information. Finally the engineer preparing the feasibility analysis should make recommendations based upon management's guidelines.

The development of the information required for Table 9-27 was discussed previously in Sec. 9 with the exception of marketing information. An important document for a feasibility analysis is the marketing data so that the latest income projections can be included for management's consideration. As a minimum, the tabulation of sales volume, sales prices, and market share both domestically and globally should be included. Table 9-28 shows a sample of such marketing information.

Other templates may be prepared for total capital investment, working capital, total product expense, general overhead expense, and cash flow. Table 9-29 may be used to organize cash flow data by showing investment, operating expenses, cash flow, and cumulative cash flow.

OTHER ECONOMIC TOPICS

COMPARISON OF ALTERNATIVE INVESTMENTS

Engineers are often confronted with making choices between alternative equipment, designs, procedures, plans or methods. The courses of action require different amounts of capital and different operating expenses. Some basic concepts must be considered before attempting to use mathematical methods for a solution. It is necessary to clearly define the alternatives and their merits. Flow of money takes the form of expenditures or income. Savings from operations are considered as income or as a reduction in operating expenses. Income taxes and inflation as well as a reasonable return on the investment must be included. Money spent is negative and money earned or saved is positive.

Expenditures are of two kinds; instantaneous like land, working capital and capital recovery or uniformly continuous for plant investment, operating expenses, etc. A methodology involving after-tax cash flow is developed to reduce all the above to a manageable format.

In an earlier part of this section, after-tax cash flow was defined as

$$\text{CF} = (R - C - D)(1 - t) + D \tag{9-17}$$

where CF = after-tax cash flow
D = depreciation
t = tax rate
S = sales or revenue
C = cash operating expenses (COE)

For the situation in which each case will produce the same revenue or the same benefit, R will equal 0. Rearranging Eq. (9-17) algebraically yields

$$\text{CF} = (t)(D) + (1 - t)(S - C) \tag{9-17a}$$

or

$$\text{CF} = (t)(D) + (1 - t)(-C) \tag{9-17b}$$

$$\text{CF} = (t)(D) - (1 - t)(C) \tag{9-17c}$$

This expression is applied to each alternative. [*Note:* As mentioned under cash flow, the first term in the above equations, $(t)(D)$, is the result of an algebraic rearrangement, and no other significance should be assumed.]

Several methods are available for determining the choice among alternatives:

Net present worth
Rate of return
Capitalized cost
Cash flow
Uniform annual cost

Humphreys in Jelen and Black, *Cost and Optimization Engineering* (1991), has shown that each of these methods would result in the same decision, but the numerical results will differ.

Net Present Worth Method The NPW method allows the conversion of all money flows to be discounted to the present time. Appropriate interest factors are applied depending on how and when the cash flow enters a venture. They may be instantaneous, as in the purchase of capital equipment, or uniform, as in operating expenses. The alternative with the more positive NPW is the one to be preferred. In some instances, the alternatives may have different lives so the cost analysis must be for the least common multiple number of years. For example, if alternative A has a 2-year life and alternative B has a 3-year life, then 6 years is the least common multiple. The rate of return, capitalized cost, cash flow, and uniform annual cost methods avoid this complication.

Rate of return and capitalized cost methods are discussed at length in Humphreys (1991).

Cash Flow Method Cash flows for each case are determined, and the case that generates the greater cash flow is the preferred one.

Uniform Annual Cost (UAC) Method In the uniform annual cost method, the cost is determined over the entire estimated project life. The least common multiple does not have to be calculated, as in the NPW method. This is the advantage of the UAC method; however, the result obtained by this method is more meaningful than the results obtained by other methods.

The UAC method begins with a calculation for each alternative. If discrete interest is used, the annual cost C is found by multiplying the present worth P by the appropriate discrete interest factor, found in Table 9-21, for the number of years n and the interest rate i. If continuous interest is preferred, the UAC equation is

$$\text{UAC} = \frac{\text{NPW}}{(\text{years of life})\ (\text{continuous interest factor})} \tag{9-23}$$

The continuous interest factor may be found from continuous interest equations or from the continuous interest table, Table 9-30. In this table time zero is the present, and all cash flows are discounted back to the present. Note that there are three sections to this table, depending on the cash flow: uniform, instantaneous, or declining uniformly to zero. One enters the table with the argument $R \times T$, where R is the interest rate expressed as a whole number and T is the time in years to obtain a factor. This factor is then used to calculate the present worth of the cash flow item. All cash flows are summed algebraically, giving the net present worth which is substituted in Eq. (9-23). This procedure is followed for both alternatives, and the alternative that yields the more positive UAC (or the least negative) value is the preferred alternative. In Eq. (9-23) the "factor" is always the uniform factor that annualizes all the various cash flows.

This method of comparing alternatives is demonstrated in Example 10.

Example 10: Choice among Alternatives Two filters are considered for installation in a process to remove solids from a liquid discharge stream to meet environmental requirements. The equipment is to be depreciated over a 7-year period by the straight-line method. The income tax rate is 35 percent, and 15 percent continuous interest is to be used. Assume that the service life is 7 years and there is no capital recovery. Data for the two systems are as follows:

System	B	C
Fixed investment	$18,000	$30,000
Annual operating expenses	14,200	4,800

Which alternative is preferred?

Solution:
System B:

Year	Item	Cash flow, $	Factor	PW, $
0	Investment	−18,000	1.0	−18,000
0–7	Contribution to cash flow from depreciation	(0.35)(18,000)	0.6191	+3,900
0–7	Contribution to cash flow from operating expense	(1 − 0.35)(7)(14,200)	0.6191	−40,000
			NPW B	−54,100

$$\text{UAC}_\text{B} = \frac{\text{NPW}}{(\text{years of life})(\text{uniform factor})} = \frac{-\$54{,}100}{(7)(0.6191)} = -\$12{,}484$$

System C:

Year	Item	Cash flow, $	Factor	PW, $
0	Investment	−30,000	1.0	−30,000
0–7	Contribution to cash flow from depreciation	(0.35)(30,000)	0.6191	+6,500
0–10	Contribution to cash flow from operating expense	(1 − 0.35)(10)(4,800)	0.6191	−19,316
			NPW C	−42,816

$$\text{UAC}_\text{C} = \frac{\text{NPW}}{(\text{years of life})(\text{uniform factor})} = \frac{-\$42{,}816}{(10)(0.6191)} = -\$6{,}916$$

Alternative C is preferred because it has the more positive UAC.

REPLACEMENT ANALYSIS

During the lifetime of a physical asset, continuation of its use may make it a candidate for replacement. In this type of analysis, a replacement is intended to supplant a similar item performing the same service without plant or equipment expansion. In a chemical plant, *replacement* usually refers to a small part of the processing equipment such as a heat exchanger, filter, or compressor. If the replacement is required due to "physical" deterioration, there is no question of whether to replace the item, but the entire plant may be shut down if it is not replaced. The problem then becomes whether the equipment

TABLE 9-30 Factors for Continuous Discounting

$R \times T$	0	1	2	3	4	5	6	7	8	9
					Uniform					
0	1.0000	.9950	.9901	.9851	.9803	.9754	.9706	.9658	.9610	.9563
10	.9516	.9470	.9423	.9377	.9332	.9286	.9241	.9196	.9152	.9107
20	.9063	.9020	.8976	.8933	.8891	.8848	.8806	.8764	.8722	.8681
30	.8639	.8598	.8558	.8517	.8477	.8437	.8398	.8359	.8319	.8281
40	.8242	.8204	.8166	.8128	.8090	.8053	.8016	.7979	.7942	.7906
50	.7869	.7833	.7798	.7762	.7727	.7692	.7657	.7622	.7588	.7554
60	.7520	.7486	.7453	.7419	.7386	.7353	.7320	.7288	.7256	.7224
70	.7192	.7160	.7128	.7097	.7066	.7035	.7004	.6974	.6944	.6913
80	.6883	.6854	.6824	.6795	.6765	.6736	.6707	.6679	.6650	.6622
90	.6594	.6566	.6538	.6510	.6483	.6455	.6428	.6401	.6374	.6348
100	.6321	.6295	.6269	.6243	.6217	.6191	.6166	.6140	.6115	.6090
110	.6065	.6040	.6015	.5991	.5967	.5942	.5918	.5894	.5871	.5847
120	.5823	.5800	.5777	.5754	.5731	.5708	.5685	.5663	.5640	.5618
130	.5596	.5574	.5552	.5530	.5509	.5487	.5466	.5444	.5423	.5402
140	.5381	.5361	.5340	.5320	.5299	.5279	.5259	.5239	.5219	.5199
150	.5179	.5160	.5140	.5121	.5101	.5082	.5063	.5044	.5025	.5007
160	.4988	.4970	.4951	.4933	.4915	.4897	.4879	.4861	.4843	.4825
170	.4808	.4790	.4773	.4756	.4738	.4721	.4704	.4687	.4671	.4654
180	.4637	.4621	.4604	.4588	.4572	.4555	.4539	.4523	.4507	.4492
190	.4476	.4460	.4445	.4429	.4414	.4399	.4383	.4368	.4353	.4338
200	.4323	.4309	.4294	.4279	.4265	.4250	.4236	.4221	.4207	.4193
210	.4179	.4165	.4151	.4137	.4123	.4109	.4096	.4082	.4069	.4055
220	.4042	.4029	.4015	.4002	.3989	.3976	.3963	.3950	.3937	.3925
230	.3912	.3899	.3887	.3874	.3862	.3849	.3837	.3825	.3813	.3801
240	.3789	.3777	.3765	.3753	.3741	.3729	.3718	.3706	.3695	.3683
250	.3672	.3660	.3649	.3638	.3627	.3615	.3604	.3593	.3582	.3571
260	.3560	.3550	.3539	.3528	.3518	.3507	.3496	.3486	.3476	.3465
270	.3455	.3445	.3434	.3424	.3414	.3404	.3394	.3384	.3374	.3364
280	.3354	.3344	.3335	.3325	.3315	.3306	.3296	.3287	.3277	.3268
290	.3259	.3249	.3240	.3231	.3222	.3212	.3203	.3194	.3185	.3176
300	.3167	.3158	.3150	.3141	.3132	.3123	.3115	.3106	.3098	.3089
310	.3080	.3072	.3064	.3055	.3047	.3039	.3030	.3022	.3014	.3006
320	.2998	.2990	.2982	.2974	.2966	.2958	.2950	.2942	.2934	.2926
330	.2919	.2911	.2903	.2896	.2888	.2880	.2873	.2865	.2858	.2850
340	.2843	.2836	.2828	.2821	.2814	.2807	.2799	.2792	.2785	.2778
350	.2771	.2764	.2757	.2750	.2743	.2736	.2729	.2722	.2715	.2709
360	.2702	.2695	.2688	.2682	.2675	.2669	.2662	.2655	.2649	.2642
370	.2636	.2629	.2623	.2617	.2610	.2604	.2598	.2591	.2585	.2579
380	.2573	.2567	.2560	.2554	.2548	.2542	.2536	.2530	.2524	.2518
390	.2512	.2506	.2500	.2495	.2489	.2483	.2477	.2471	.2466	.2460
400	.2454	.2449	.2443	.2437	.2432	.2426	.2421	.2415	.2410	.2404
					Instantaneous					
0	1.0000	.9900	.9802	.9704	.9608	.9512	.9418	.9324	.9231	.9139
10	.9048	.8958	.8869	.8781	.8694	.8607	.8521	.8437	.8353	.8270
20	.8187	.8106	.8025	.7945	.7866	.7788	.7711	.7634	.7558	.7483
30	.7408	.7334	.7261	.7189	.7118	.7047	.6977	.6907	.6839	.6771
40	.6703	.6637	.6570	.6505	.6440	.6376	.6313	.6250	.6188	.6126
50	.6065	.6005	.5945	.5886	.5827	.5769	.5712	.5655	.5599	.5543
60	.5488	.5434	.5379	.5326	.5273	.5220	.5169	.5117	.5066	.5016
70	.4966	.4916	.4868	.4819	.4771	.4724	.4677	.4630	.4584	.4538
80	.4493	.4449	.4404	.4360	.4317	.4274	.4232	.4190	.4148	.4107
90	.4066	.4025	.3985	.3946	.3906	.3867	.3829	.3791	.3753	.3716
100	.3679	.3642	.3606	.3570	.3535	.3499	.3465	.3430	.3396	3362
110	.3329	.3296	.3263	.3230	.3198	.3166	.3135	.3104	.3073	.3042
120	.3012	.2982	.2952	.2923	.2894	.2865	.2837	.2808	.2780	.2753
130	.2725	.2698	.2671	.2645	.2618	.2592	.2567	.2541	.2516	.2491
140	.2466	.2441	.2417	.2393	.2369	.2346	.2322	.2299	.2276	.2254
150	.2231	.2209	.2187	.2165	.2144	.2122	.2101	.2080	.2060	.2039
160	.2019	.1999	.1979	.1959	.1940	.1920	.1901	.1882	.1864	.1845
170	.1827	.1809	.1791	.1773	.1755	.1738	.1720	.1703	.1686	.1670
180	.1653	.1637	.1620	.1604	.1588	.1572	.1557	.1541	.1526	.1511
190	.1496	.1481	.1466	.1451	.1437	.1423	.1409	.1395	.1381	.1367
200	.1353	.1340	.1327	.1313	.1300	.1287	.1275	.1262	.1249	.1237
210	.1225	.1212	.1200	.1188	.1177	.1165	.1153	.1142	.1130	.1119
220	.1108	.1097	.1086	.1075	.1065	.1054	.1044	.1033	.1023	.1013
230	.1003	.0993	.0983	.0973	.0963	.0954	.0944	.0935	.0926	.0916
240	.0907	.0898	.0889	.0880	.0872	.0863	.0854	.0846	.0837	.0829
250	.0821	.0813	.0805	.0797	.0789	.0781	.0773	.0765	.0758	.0750
260	.0743	.0735	.0728	.0721	.0714	.0707	.0699	.0693	.0686	.0679
270	.0672	.0665	.0659	.0652	.0646	.0639	.0633	.0627	.0620	.0614
280	.0608	.0602	.0596	.0590	.0584	.0578	.0573	.0567	.0561	.0556
290	.0550	.0545	.0539	.0534	.0529	.0523	.0518	.0513	.0508	.0503
300	.0498	.0493	.0488	.0483	.0478	.0474	.0469	.0464	.0460	.0455
310	.0450	.0446	.0442	.0437	.0433	.0429	.0424	.0420	.0416	.0412

TABLE 9-30 Factors for Continuous Discounting *(Concluded)*

$R \times T$	0	1	2	3	4	5	6	7	8	9
					Instantaneous					
320	.0408	.0404	.0400	.0396	.0392	.0388	.0384	.0380	.0376	.0373
330	.0369	.0365	.0362	.0358	.0354	.0351	.0347	.0344	.0340	.0337
340	.0334	.0330	.0327	.0324	.0321	.0317	.0314	.0311	.0308	.0305
350	.0302	.0299	.0296	.0293	.0290	.0287	.0284	.0282	.0279	.0276
360	.0273	.0271	.0268	.0265	.0263	.0260	.0257	.0255	.0252	.0250
370	.0247	.0245	.0242	.0240	.0238	.0235	.0233	.0231	.0228	.0226
380	.0224	.0221	.0219	.0217	.0215	.0213	.0211	.0209	.0207	.0204
390	.0202	.0200	.0198	.0196	.0194	.0193	.0191	.0189	.0187	.0185
400	.0183	.0181	.0180	.0178	.0176	.0174	.0172	.0171	.0169	.0167
					Declining Uniformly to 0					
0	1.0000	.9968	.9934	.9902	.9867	.9836	.9803	.9771	.9739	.9707
10	.9675	.9643	.9612	.9580	.9549	.9518	.9487	.9457	.9426	.9396
20	.9365	.9335	.9305	.9275	.9246	.9216	.9187	.9158	.9129	.9100
30	.9071	.9042	.9013	.8985	.8957	.8929	.8901	.8873	.8845	.8817
40	.8790	.8763	.8735	.8708	.8681	.8655	.8628	.8601	.8575	.8549
50	.8522	.8496	.8470	.8445	.8419	.8393	.8368	.8343	.8317	.8292
60	.8267	.8242	.8218	.8193	.8169	.8144	.8120	.8096	.8072	.8048
70	.8024	.8000	.7977	.7953	.7930	.7906	.7883	.7860	.7837	.7814
80	.7792	.7769	.7746	.7724	.7702	.7679	.7657	.7635	.7613	.7591
90	.7570	.7548	.7526	.7505	.7484	.7462	.7441	.7420	.7399	.7378
100	.7358	.7337	.7316	.7296	.7275	.7255	.7235	.7215	.7195	.7175
110	.7155	.7135	.7115	.7096	.7076	.7057	.7038	.7018	.6999	.6980
120	.6961	.6942	.6923	.6905	.6886	.6867	.6849	.6830	.6812	.6794
130	.6776	.6757	.6739	.6721	.6704	.6686	.6668	.6650	.6633	.6615
140	.6598	.6581	.6563	.6546	.6529	.6512	.6495	.6478	.6461	.6445
150	.6428	.6411	.6395	.6378	.6362	.6345	.6329	.6313	.6297	.6281
160	.6265	.6249	.6233	.6217	.6202	.6186	.6170	.6155	.6139	.6124
170	.6109	.6093	.6078	.6063	.6048	.6033	.6018	.6003	.5988	.5973
180	.5959	.5944	.5929	.5915	.5900	.5886	.5872	.5857	.5843	.5829
190	.5815	.5801	.5787	.5773	.5759	.5745	.5731	.5718	.5704	.5690
200	.5677	.5663	.5650	.5636	.5623	.5610	.5596	.5583	.5570	.5557
210	.5544	.5531	.5518	.5505	.5492	.5480	.5467	.5454	.5442	.5429
220	.5417	.5404	.5392	.5379	.5367	.5355	.5342	.5330	.5318	.5306
230	.5294	.5282	.5270	.5258	.5246	.5234	.5223	.5211	.5199	.5188
240	.5176	.5165	.5153	.5142	.5130	.5119	.5108	.5096	.5085	.5074
250	.5063	.5052	.5041	.5029	.5018	.5008	.4997	.4986	.4975	.4964
260	.4953	.4943	.4932	.4922	.4911	.4900	.4890	.4879	.4869	.4859
270	.4848	.4838	.4828	.4818	.4807	.4797	.4787	.4777	.4767	.4757
280	.4747	.4737	.4727	.4717	.4707	.4698	.4688	.4678	.4669	.4659
290	.4649	.4640	.4630	.4621	.4611	.4602	.4592	.4583	.4574	.4564
300	.4555	.4546	.4537	.4527	.4518	.4509	.4500	.4491	.4482	.4473
310	.4464	.4455	.4446	.4438	.4429	.4420	.4411	.4402	.4394	.4385
320	.4376	.4368	.4359	.4351	.4342	.4334	.4325	.4317	.4308	.4300
330	.4292	.4283	.4275	.4267	.4259	.4251	.4242	.4234	.4226	.4218
340	.4210	.4202	.4194	.4186	.4178	.4170	.4162	.4154	.4147	.4139
350	.4131	.4123	.4115	.4108	.4100	.4092	.4085	.4077	.4070	.4062
360	.4055	.4047	.4040	.4032	.4025	.4017	.4010	.4003	.3995	.3988
370	.3981	.3973	.3966	.3959	.3952	.3945	.3937	.3930	.3923	.3916
380	.3909	.3902	.3895	.3888	.3881	.3874	.3867	.3860	.3854	.3847
390	.3840	.3833	.3826	.3820	.3813	.3806	.3799	.3793	.3786	.3779
400	.3773	.3766	.3760	.3753	.3747	.3740	.3734	.3727	.3721	.3714

SOURCE: Adapted and abridged from Couper, 2003.

should be replaced like for like or whether an alternative should be chosen that may be different in cost and/or efficiency. If the replacement is due to technical obsolescence, the timing of the replacement may be important, especially if a plant expansion may be imminent in the near future. Whatever the situation, the replaced item should not present a bottleneck to the processing. The engineer should understand replacement theory to determine if alternative equipment is adequate for the job but with different costs and timing.

Certain terminology has been developed to identify the equipment under consideration. The item in place is called the *defender*, and the candidate for replacement is called the *challenger*. This terminology and methodology was reported by E. L. Grant and W. G. Ireson in *Engineering Economy*, Wiley, New York, 1950. To apply this method, there are certain rules. The value of the defender asset is a sunk cost and is irrelevant except insofar as it affects cash flow from depreciation for the rest of its life and a tax credit for the book loss if it is replaced sooner than its depreciation life. A capital cost for the defender is the net capital recovery forgone and the tax credit from the book loss of the defender asset that was not realized. The UAC method will be used and will be computed for each case, using the time period most favorable to each. For the defender it is 1 year and for the challenger it is the full economic life. The UAC for the challenger is handled in the same manner as in the comparison of alternatives. The method is demonstrated in Example 11.

Example 11: Replacement Analysis A 3-year-old reciprocating compressor is being considered for replacement. Its original cost was \$150,000, and it was being depreciated over a 7-year period by the straight-line method. If it is replaced now, the net proceeds from its sale are \$50,000, and it is believed that 1 year from now they will be \$35,000. A new centrifugal compressor can be installed for \$160,000, which would save the company \$2000 per year in operating expenses for the 10-year life. At the end of the 10th year, its net proceeds are estimated to be zero. The 7-year depreciation applies also to the centrifugal compressor. A 35 percent tax rate may be assumed. The company requires a 15 percent after-tax return on an investment of this type. Should the present compressor be replaced now?

Solution: The UAC method will be used as a basis for comparison. It is assumed that all money flows are continuous, and continuous interest will be used.

Defender case: The basis for this unit will be 1 year. If it is not replaced now, the rules listed above indicate that there is an equivalent of a capital cost for two benefits forgone (given up). They are

1. Net proceeds now at 3 years of $50,000
2. Tax credit for the loss not realized

Thus net loss forgone = book value at the end of 3 years minus net capital recovery, or

$$NLF_3 = BV_3 - NCR_3$$

where NLF_3 = net loss forgone at end of 3 years
BV_3 = book value at end of 3 years
NCR_3 = net capital recovery at end of 3 years

$$NLF_3 = \$150{,}000\left(1 - \frac{1}{7}\right) - \$50{,}000 = \$35{,}714$$

$$\text{Depreciation for 4th year} = \$150{,}000\left(\frac{1}{7}\right) = \$21{,}429$$

$$NLF_4 = BV_4 - NCR_4 = \$150{,}000\left(1 - \frac{4}{7}\right) - \$35{,}000 = \$29{,}286$$

Year	Item	Cash flow, $	Factor	PW, $
At 0[a]	Tax credit for net loss forgone	(0.35)(−35,714)	1.0	−12,500
At 0[a]	Net cash recovery forgone	−50,000	1.0	−50,000
0–1	Contribution to CF from depreciation	(0.35)(−21,429)	0.9286[b]	6,965
0–1	Contribution to CF from operting expense	(1 − 0.35)(−15,000)	0.9286[b]	−9,054
End 1	Tax credit for net loss	(0.35)(29,286)	0.8607[c]	8,822
End 1	Net cash	35,000	0.8607[c]	30,125
			NPW	−25,642

[a]For the defender case, 0 year is the end of the third year.
[b]From Table 9-30, uniform section, for the argument $R \times T = 15 \times 1 = 15$.
[c]From Table 9-30, instantaneous section, for the argument $R \times T = 15 \times 1 = 15$.

$$UAC = \frac{NPW}{(\text{years of life})(\text{uniform factor})} = \frac{-\$25{,}642}{(1)(0.9286)} = -\$27{,}614$$

Challenger case:

Year	Item	Cash flow, $	Factor	PW, $
0	First cost	−160,000	1.0	−160,000
0–7	Contribution to CF from depreciation	(0.35)(160,000)	0.6191[d]	56,000
0–10	Contribution	(1 − 0.35)(10)(−13,000)	0.5179[e]	−43,763
			NPW	−147,763

[d]From Table 9-30, uniform section, $R \times T = 15 \times 7 = 105$.
[e]From Table 9-30, uniform section, $R \times T = 15 \times 10 = 150$.

$$UAC = \frac{NPW}{(\text{years of life})(\text{uniform factor})} = \frac{-\$147{,}763}{(10)(0.5179)} = -\$28{,}531$$

The UAC for the defender case is less negative (more positive) than that for the challenger case; therefore, the defender should not be replaced now. But there will be a time in the near future when the defender should be replaced, as maintenance and deterioration will increase.

OPPORTUNITY COST

Opportunity cost refers to the cost or value that is forgone or given up because a proposed investment is undertaken, often used as a base case. Perhaps the term should be *lost opportunity*. For example, the profit from production in obsolete facilities is an opportunity cost of replacing them with more efficient ones. In cost analysis on investments, an incremental approach is often used, and if it is applied properly, the correct cost analysis will result.

ECONOMIC BALANCE

An engineering cost analysis can be used to find either a minimum cost or a maximum profit for a venture. This analysis is called an *economic balance* since it involves the balancing of economic factors to determine optimum design or optimum operating conditions. Such an analysis involves *engineering tradeoffs*. It may be more beneficial to invest more capital to reduce operating expenses or, conversely, incur more operating expenses without the addition of costly capital. An economic balance, then, is a study of all costs, expenses, revenues, and savings that pertain to an operation or perhaps an equipment item size. In this presentation, certain terminology is used; e.g., the term *cost* refers to a one-time purchase of capital equipment. A recurring expense is called an *operating expense,* such as utilities, labor, and maintenance. All costs and operating expenses are related to an arbitrarily designated *controllable* variable such as heat exchanger area, thickness of insulation, or number of units.

There are certain practical considerations that must be recognized in attempting to find the best or optimum condition. Occasionally, a solution may lead to a result for which industrial equipment is not available in the optimum size; therefore, engineering judgment must be exercised. A smaller-diameter pipe might lead to a higher pumping expense but lower pipe costs while a larger-diameter pipe would result in lower pumping expense but higher capital cost. The engineer then encounters an engineering tradeoff that must be resolved.

The essential elements of an economic balance are

- Fixed and variable expenses
- An allowance for depreciation
- An acceptable return on the investment

A total expense equation is

$$TE = FE + VE \tag{9-24}$$

where TE = total expenses
FE = fixed expenses
VE = variable expenses

The guidelines for solving a single-variable economic balance consist of the following steps:

1. Determine all expenses that vary as the controllable variable changes and that need to be considered in the balance.
2. Determine whether any operating limitations exist, such as pressure drop in columns where flooding occurs or limiting heads for pipelines for gravity flow.
3. Mathematically express the expenses as a function of the variables related to the equipment; otherwise use variables that define the operation such as temperature, pressure, and concentration. The final expression should include all pertinent expenses, eliminating those that are not significant. Frequently only one variable is used.
4. Ascertain if the optimum size must be one of a number of discrete sizes commercially available or whether it can be any size.
5. Solve the total expense equation either analytically or graphically.

The solution to Eq. (9-24) may be found analytically or graphically. In the analytical method, the fixed and variable expenses are related in equation format to the controllable variable. Example 12 is an engineering balance example of this method. Equation (9-24) can be differentiated with respect to the controllable variable and that result set equal to zero to find the optimum condition. A graphical method involves determining the fixed and variable expenses for a range of equipment sizes. A plot of TE, FE, and VE as a function of the controllable sizes yields a plot identifying the optimum. Figure 9-18 is a graphical solution to the TE equation. Should the optimum result fall between two commercial sizes, then the engineer must exercise judgment in the tradeoff.

Example 12: Optimum Number of Evaporator Effects Determine the number of evaporator effects for the minimum total annual operating

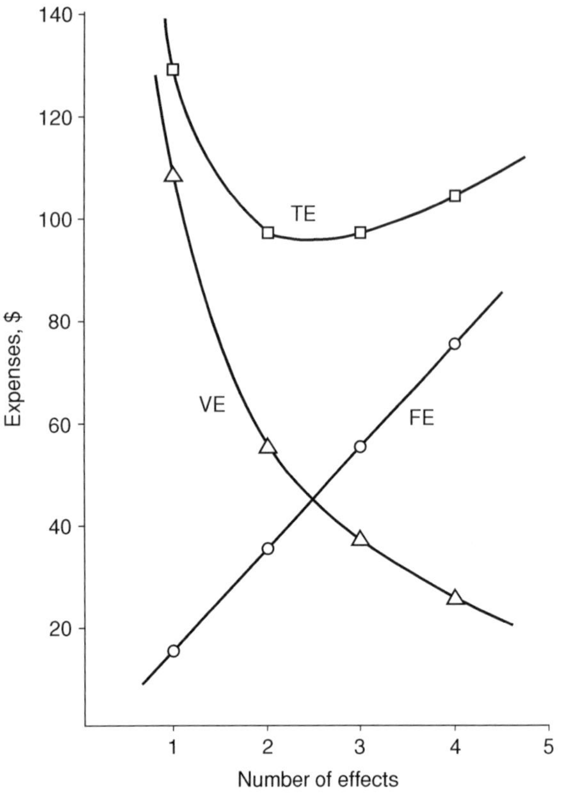

FIG. 9-18 Optimum number of evaporator effects in Example 12.

expense of a small evaporator system to concentrate a colloidal suspension. Steam costs \$3.00 per million Btu (MBtu), and each pound of steam will evaporate 0.8*N* lb water, with *N* being the number of effects. The total capital cost of each effect is \$40,000 and has an estimated life of 10 years. The annual maintenance expense is 10 percent of the capital cost. Labor and other expenses not mentioned may be considered to be independent of the number of effects. The system will operate 300 days/yr with 100,000,000 Btu/day evaporator duty. Depreciation on the equipment is by 7-year straight-line method, and the tax rate is 35 percent. Annual net profit after taxes on the investment must be 15 percent for the installed equipment.

Basis: 1 year of operation, duty 10^8 Btu/day. Let *N* equal the number of effects.
Solution:
Analytical method:

Annual fixed expenses = FE

$$\text{FE} = \$40{,}000N\left(0.143 + 0.100 + \frac{0.15}{1-0.35}\right) = \$18{,}960N$$

dep. maint. profit (return)

Annual variable expenses VE in this example are essentially the steam expenses, and all other variable expenses do not enter into the equation. Therefore

$$\text{VE} = 10^8 \text{ Btu/day} \times 300 \text{ day/yr} \times \$3.00/10^6 \text{ Btu} \times 1/(0.8N) = \frac{\$112{,}500}{N}$$

$$\text{TE} = \text{FE} + \text{VE} = \$18{,}960N + \frac{\$112{,}500}{N}$$

Take the derivative of TE with respect to *N* and set the result equal to zero.

$$\frac{d\,(\text{TE})}{dN} = 18{,}960 - \frac{112{,}500}{N^2} = 0 \quad N = 2.44 \text{ effects}$$

Therefore 3 effects are required.
Graphical method: Assume the number of effects and calculate the fixed, variable, and total expenses for effect.

N	18,960*N*	112,500/*N*	TE
1	\$18,960	\$112,500	\$131,460
2	37,920	56,250	94,210
3	56,880	37,500	94,380
4	75,840	28,125	103,965

The minimum TE occurs between 2 and 3 effects, but 3 effects are recommended to evaporate the colloidal solution. Figure 9-18 is a plot of the graphical solution.

In another example involving the reclaiming of a product using an evaporator and a dryer in series, the product is pumped from the discharge of the evaporator to the dryer. See Fig. 9-19. An economic analysis indicated that a 55 percent slurry is optimum, but perhaps such a slurry is too thick to be pumped. Therefore, engineering judgment must be exercised. The slurry was pumped not at 55 percent but at a lesser concentration of 50 percent, although it was not optimum.

When more than one controllable variable affects the economic balance, the solution approach is essentially the same as that for the single-variable case, but determining the optimum is tedious.

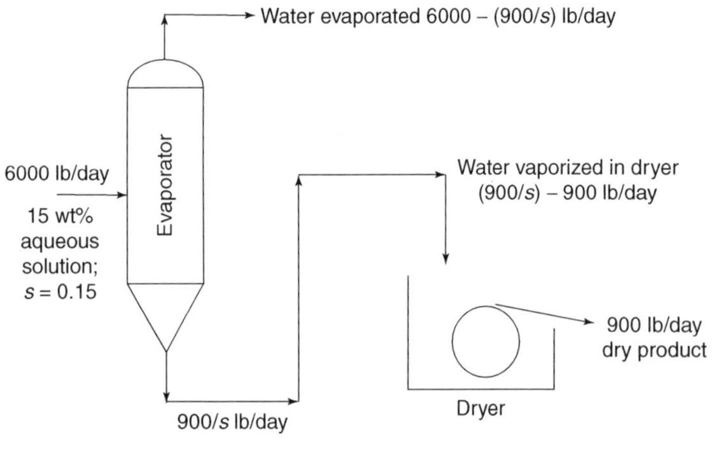

FIG. 9-19 Evaporator–dryer system. (*Source: Adapted from Couper, 2003.*)

INTERACTIVE SYSTEMS

If the TE does not pass through a minimum or maximum, but continues to decline or to increase with the number of equipment items or equipment size, the next step is to look at the flow sheet for equipment upstream or downstream from the selected item. It may be necessary to group two or more items and treat them as one in the analysis. Such a system is said to be *interactive*, since more than one item affects the optimum results. An example of such an interactive system is the removal of nitrogen from helium in a natural gas stream. Carbon adsorption is a method for removing nitrogen, but compressors are also required since this is a high-pressure process. If one attempts to find the optimum operating pressure, optimizing on compressor pressure will not result in an optimum condition; and conversely, optimizing on the size of the carbon bed will not yield an optimum. This is an example of an interactive system. Therefore, to find the optimum pressure, both the size of the carbon bed and the compressor pressure must be considered together.

CAPITAL PROJECT EXECUTION AND ANALYSIS

The demands made on business organizations with the arrival of global free trade have made sophisticated management of capital projects, for the purposes of minimizing capital costs and maximizing project profitability, a necessity. Two elements of such sophisticated project management practice are *front-end loading* (*FEL*) and *value-improving practices* (*VIPs*). These two management practices are, and must be, closely integrated activities, as seen in Fig. 9-20. As shown, they are performed during the early stages of a project life cycle, when they can be, and are, effective at influencing a project's profitability. However, they have very different characteristics, as are detailed in the following sections. Properly performed together, FEL and VIPs maximize project profitability by ensuring that all matters that influence project profitability are considered in the most productive manner and at the most optimal time.

FRONT-END LOADING

GENERAL REFERENCES: Porter, James B., E. I. DuPont de Nemours and Company, DuPont's Role in Capital Projects, Proceedings of Government/Industry Forum: The Owner's Role in Project Management and Pre-project Planning, 2002. Smith, C. C., Improved Project Definition Insures Value-added Performance—Part 1, *Hydrocarbon Processing*, August 2000, pp. 95–99. KBR Front-end Loading Program, data compiled from selected large projects from 1993 through 2003, http://www.halliburton.com/kbr/index.jsp. Merrow, E. W., Independent Project Analysis, Inc., 32d Annual Engineering & Construction Contracting Conference, Sept. 28–29, 2000. Merrow, E. W., Independent Project Analysis, Inc., 30th Annual Engineering & Construction Contracting Conference, September 1998. HPI Impact, *Hydrocarbon Processing*, August 2002, p. 23, data obtained from Independent Project Analysis, Inc. PDRI: Project Definition Rating Index— Industrial Projects, p. 5, Construction Industry Institute (CII), University of Texas at Austin, http://construction-institute.org, July 1996. Merrow, E. W., Independent Project Analysis, Inc., 32d Annual Engineering & Construction Contracting Conference, Sept. 28–29, 2000.

Introduction Front-end loading (FEL) is the process by which a company develops a detailed definition of the scope of a capital project that meets corporate business objectives. The term *front-end loading* was first coined by the DuPont company in 1987 and has been used throughout the chemical, refining, and oil and gas industries ever since (Porter, James B., E. I. DuPont de Nemours and

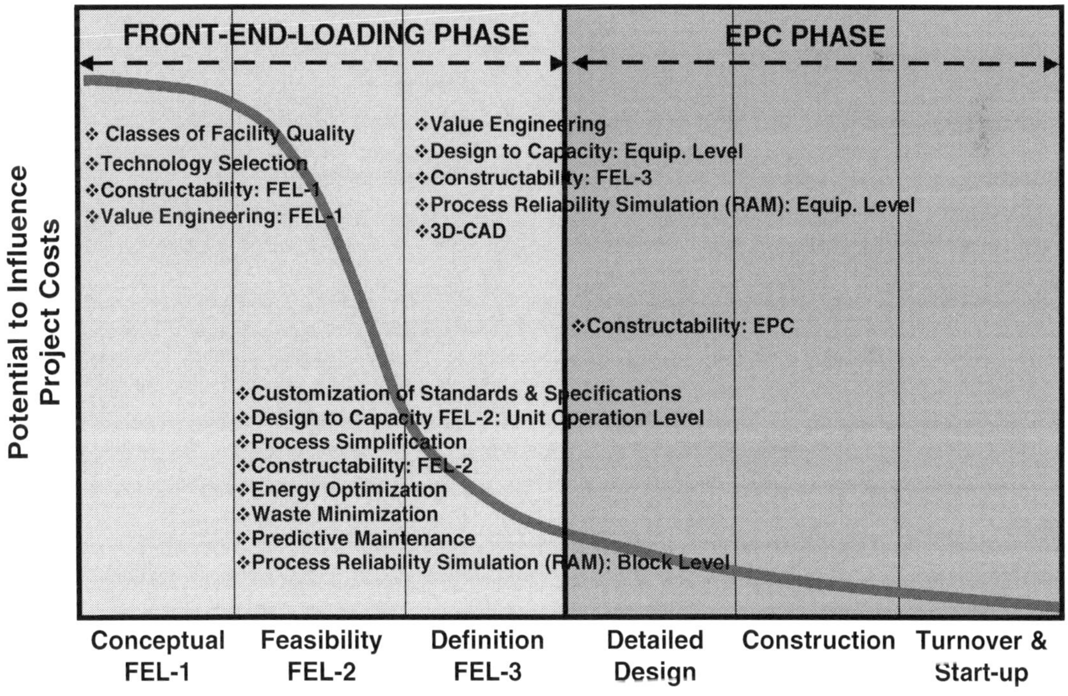

FIG. 9-20 Front-end loading and the implementation of value-improving practices.

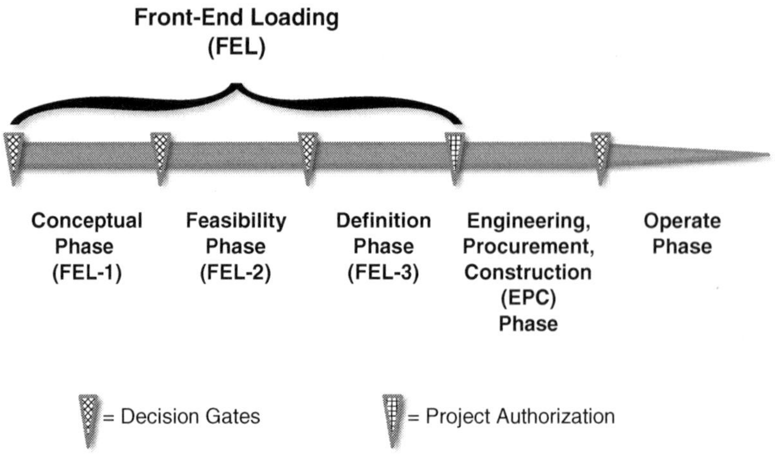

FIG. 9-21 FEL in the capital project life cycle.

Company). The product of the FEL process is a design-basis package of customized information used to support the production of detailed engineering design documents. Completion of the FEL design-basis package typically coincides with project AFE (Authorization for Expenditure) or project authorization. Project authorization is that point in the project life cycle where the owner organization commits the majority of the project's capital investment and contracts.

FEL starts when an idea for a project is first conceived by a research and development group, project engineering group, plant group, or business unit. FEL activity continues until the project is authorized. After initial conception of an idea, organized interaction is required among the various project stakeholders to assemble the project design-basis package for subsequent authorization.

Within the FEL phases, decision points are formally established by the operating company authorizing the initiation of a capital project development effort. These formal decision gates allow for continuity across the enterprise for authorization of additional funding for the next phase of engineering and project definition. Figure 9-21 illustrates the typical decision gates or stage gates for capital projects.

When the level of project definition is sufficient to support a definitive cost estimate for both the entire project and its projected rate of return, major project funding is authorized for expenditure. This is the conclusion of the FEL process and any significant involvement of the process design engineer. Not until the conclusion of the engineering, procurement, and construction (EPC) phase does the process engineer again become involved. At this time, commissioning and start-up become the focus where the validation occurs for all that was done in the FEL phases many months earlier.

Differing terminology used by companies, engineers, and project management teams is often a point of confusion. Most people seem to think they know what all the terms mean. This is never the case. Confirmation of which terminology will be used by all involved in the project is a must. Nearly every operating company and engineering contractor use differing terminology. FEL terminology is often misunderstood and further confused by differing references to which FEL phase the project is actually in. Figure 9-22 provides some idea of the differing terminology for each project phase used by only a few major oil and gas and chemical companies today. These terms change periodically, so diligence in confirming such terminology is a key task for the process engineers to finish, before beginning their work.

The influence of changes on capital projects is considerably affected by when those changes occur. The earlier a change is considered and incorporated into the project scope, the greater its potential influence on the project's profitability and the greater the ease of incorporating the change. This means that late changes (e.g., in the EPC phase) are far more expensive to implement and are considered very undesirable. Late changes which are potentially advantageous are often not implemented because the cost to implement the change exceeds the benefits of doing so. Conversely, the cost to implement a change at the earlier phases of the project is far lower than making the same change after detailed engineering is underway.

Figure 9-23 shows how quickly this influence curve changes as the typical project progresses (Smith, C. C., Improved Project Definition Insures Value-added Performance—Part 1). This is why proactively seeking changes during FEL is far more advantageous to profitability than is allowing those needed changes to be "discovered" during later project phases. This also means that potentially beneficial changes (value improvements) must be sought during FEL, or else they stand a good chance of not being cost-effective to implement during the EPC phase. This is also why seeking operations, maintenance, and construction experience during FEL offers significant profitability advantages over practices which bring such experience onto the project team following FEL.

Characteristics of FEL Front-end loading is a specialized and adaptable work process. This work process translates financial and marketing opportunities to a technical reality in the form of a capital project. It is particularly important that the project be defined in sufficient detail by the engineering deliverables, which are generated by the FEL work process, prior to the point where major funds are authorized. In this manner, overall project risks are identified and sufficiently mitigated to have project funding approved. To achieve this important level of definition, critical decisions must be made and adhered to. In addition, the FEL project team should proactively seek value improvement alternatives and challenge the project premises, scope, and design until such time as implementation of those alternatives loses their profitability and/or technical advantage. By doing so, such value improvements will not develop into costly corrections, which surface later, during the EPC phase.

Goals and Objectives of FEL The FEL work process must enable nearly constant consideration of changes as the work progresses. FEL phases must consider the long-term implications of every aspect of the design. Predictability of equipment and process system life cycle costs must always be balanced with operations and maintenance preferences, as well as the need for the project to maintain its profitability or ROI (return on investment). Additional important goals and objectives of FEL projects are as follows:

- Develop a well-defined and acceptably profitable project.
- Define the primary technical and financial drivers for capital project investment.

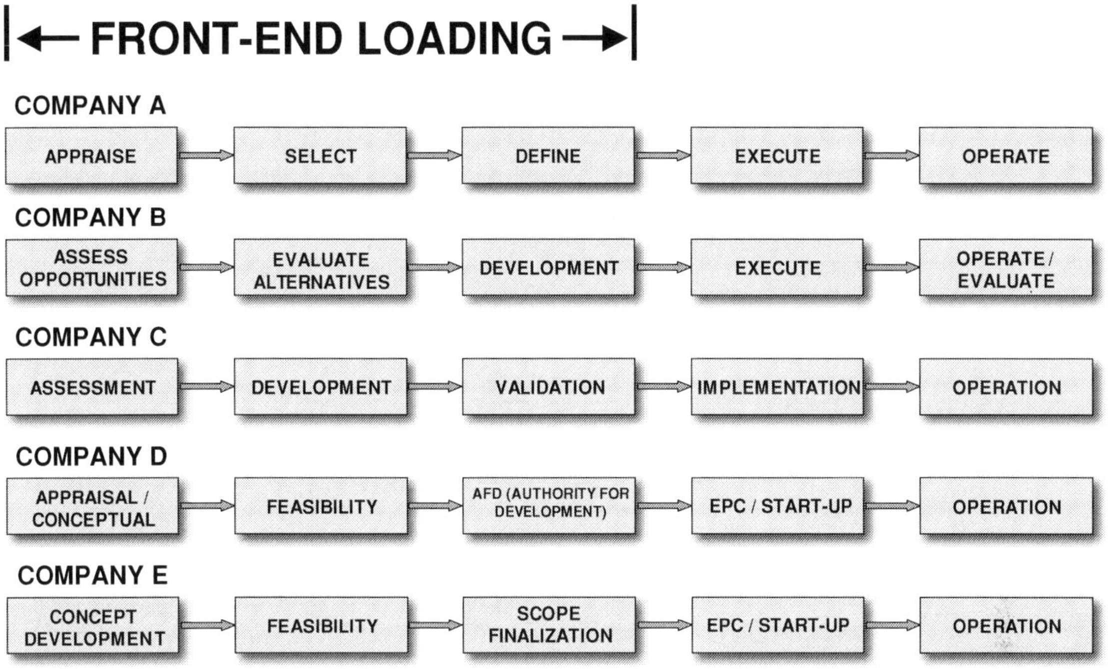

FIG. 9-22 Project life cycle terminology.

- Challenge baseline premises, and purposely seek out and evaluate alternatives and opportunities.
- Minimize changes during the EPC, turnover, and start-up phases.
- Reduce project schedule and capital cost.
- Reduce the business and project execution risk.
- Balance project technical, financial, and operational profitability drivers.

Comparison of FEL Projects with EPC Projects FEL projects are very different from EPC projects. Engineers and project managers having significant experience only with projects in the EPC phase often are unfamiliar with the significant differences between the philosophies and challenges of the FEL phase and the EPC phase of projects. One of the most important (but most subtle) aspects of FEL is the demand during FEL for more highly experienced staff and

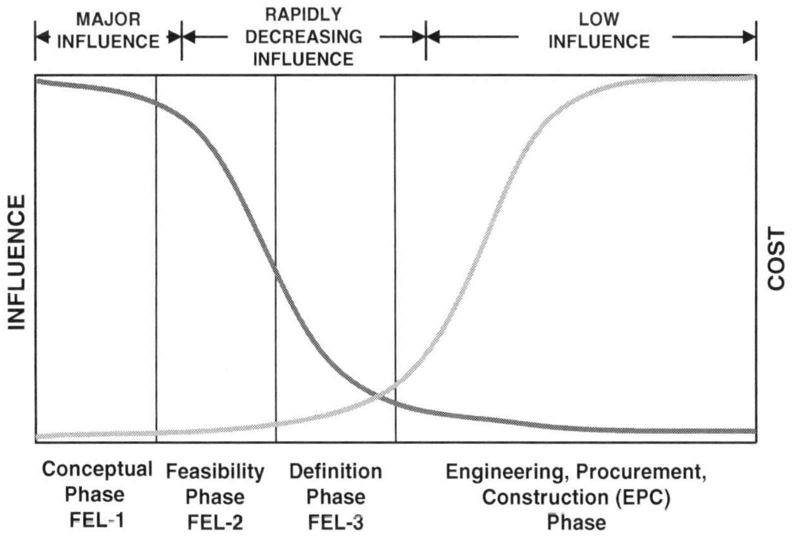

FIG. 9-23 Project life cycle cost-influence curve.

	FEL	EPC
Project State	Undefined	Defined
Changes	Actively Seeks Changes	Actively Resists Changes
Impact of Change	Low	High
Opportunity for Change	High	Low
Contract Type	Typically Reimbursable	Typically Lump Sum
Value Improvement Potential	High	Low
Client Participation	Encouraged	Discouraged
Philosophy	Information-Driven	Deliverable-Driven

FIG. 9-24 FEL projects versus EPC projects.

more sophisticated analysis tools, as compared to EPC projects which have achieved a well-defined project prior to authorization. This is so because of the need in FEL to create, analyze, and implement improvements to what many might consider a "good" design.

In spite of its relatively short duration, FEL proactively seeks to implement the best possible design. The nearly constantly changing environment requires people of many different disciplines and functions to work together to communicate effectively. A well-integrated team always seems to perform best during FEL, if FEL has well-established, informal, and personal interfaces between project groups and organizations. The following describes how the FEL phase is distinguished from the EPC phase:

- FEL proactively seeks data, resources, support, and decision making.
- Projects in the FEL phase place a higher level of importance on close and effective owner-contractor management interfaces.
- FEL demands continuous realignment of client desires and requirements with contractor needs.
- FEL requires greater development of personal relationships that result in respect and trust.
- FEL demands significantly higher frequency of feedback of owner satisfaction.
- FEL emphasizes elimination of low- or zero-value scope.
- FEL improves the capital productivity of projects by using best-available technology.
- FEL focuses on overall project profitability rather than on only cost, schedule, and workhours.
- FEL focuses almost entirely on the owner's business needs.

Figure 9-24 lists further differences between FEL and EPC projects. Understanding these many differences is very important to the process engineer, in that awareness of them, and the driving forces behind them, will prepare the chemical engineer for the challenging and rewarding environment of FEL projects.

Parameters of FEL Phases Important aspects of each phase of FEL are cost estimate accuracy, cumulative engineering hours spent, and the contingency assigned to the cost estimate. Figure 9-25 lists the typical parameters encountered industrywide (KBR Front-end Loading Program, data compiled from selected large projects from 1993 through 2003). For the capital cost estimate, each operating company may request a slightly different accuracy, which is often project-specific. What is important is the level of engineering required to support such estimating accuracy. This determination is the responsibility of both the owner and the engineering contractor. Agreement on this is critical prior to initiating project work.

The engineering hours spent during each phase of FEL vary widely between small and large projects. This is also true for those projects where new or emerging technology is being applied or where higher throughput capacities are being applied than previously commercially demonstrated. Projects such as these may require additional engineering to achieve the desired estimate accuracy and project contingency.

FEL Project Performance Characteristics Overall project performance can be enhanced by ensuring that the following characteristics are emphasized during the FEL phases.

- Methodical business and project execution planning is necessary.
- Effective integration of workforce between owner and contractor staff is important.
- Projects with an integrated management team (owner and engineering contractor) have the lowest number of design changes at any project stage.
- Engineering contractor should be brought into project in early FEL phases.
- Clear roles must exist for project team members that relate to the expertise of both owner and contractor staff.
- Effective personal communication is required between owner and contractor organizations and their project team representatives, ensuring extensive site and manufacturing input.
- Schedule and cost goals are set by integrated business and technical project team composed of owner and contractor representatives.

Figure 9-26 illustrates the benefit of good FEL performance on project costs (Merrow, E. W., Independent Project Analysis, Inc., 32d Annual Engineering & Construction Contracting Conference, Sept. 28–29, 2000). Figure 9-27 illustrates the benefit of good FEL performance on critical path schedule (Merrow, E. W., Independent

	FEL-1 (CONCEPTUAL)	FEL-2 (FEASIBILITY)	FEL-3 (DEFINITION)
Cost Estimate Accuracy	±40%	±25%	±10%
Cumulative Engineering Hours Spent	1–5%	5–15%	15–30%
Contingency	15–20%	10–15%	8–12%

FIG. 9-25 Parameters of FEL phases.

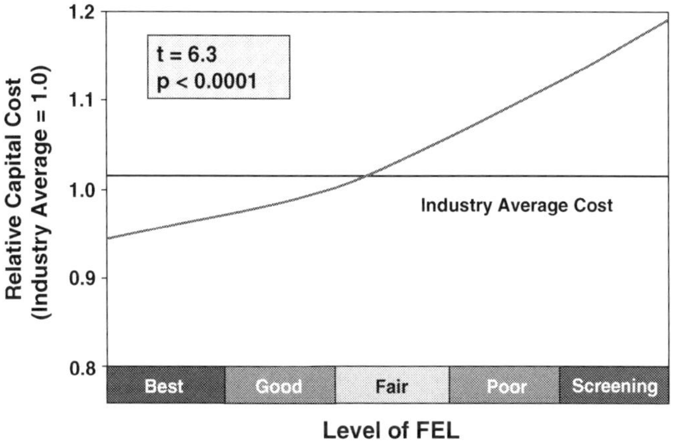

FIG. 9-26 FEL drives better cost performance.

Project Analysis, Inc., 32d Annual Engineering & Construction Contracting Conference, Sept. 28–29, 2000). IPA statistics indicate that significant project financial and schedule benefits can be realized by implementing a thorough FEL effort prior to the EPC phase.

Figure 9-28 presents the benefits of having an integrated project team during FEL on the overall project performance. This performance impacts overall project costs as well as schedule and operability. An integrated project team produces fewer late changes. This means lower capital costs, better and more predictable schedules, and a slightly better operability, as compared statistically to similar projects lacking an integrated management team (Merrow, E. W., Independent Project Analysis, Inc., 30th Annual Engineering & Construction Contracting Conference, September 1998).

In addition, project data indicate that a well-integrated FEL team can produce significantly better project performance in terms of lower capital investment, as compared to projects where FEL teams were not properly integrated. This illustrates the benefits for each engineering team member working closely together with each other team member, to produce the most profitable project results.

Although project teams, once integrated and functioning with clear roles and responsibilities, perform better, this edge can be quickly lost if key members of that team are changed. The impacts of changes of project managers to a well-integrated FEL team are shown in Fig. 9-29 (HPI Impact, *Hydrocarbon Processing*, p. 23, August 2002, data obtained from Independent Project Analysis, Inc.).

Investment in FEL for Best Project Performance The cost and schedule required to optimally complete the FEL phase of a project are always under pressure and must be justified. This is especially true for "fast-track" projects where the time pressures can be significant. The Construction Industry Institute (CII) has shown that higher levels of preproject planning (i.e., front-end loading) effort can result in significant cost and schedule savings, as seen in Fig. 9-30 (PDRI: Project Definition Rating Index—Industrial Projects, Construction Industry Institute, University of Texas at Austin, July 1996). The process engineer produces the best project performance, when he or she strives, with the entire integrated FEL project team, to define the overall project (not just the process design) as well as possible, prior to AFE.

The level of definition of a project during the FEL phases has a direct influence on the project's ultimate outcome in terms of the number and impacts of changes in the EPC phase. This level of FEL performance translates to fewer major changes in engineering, construction, and during start-up (Merrow, E. W., Independent Project

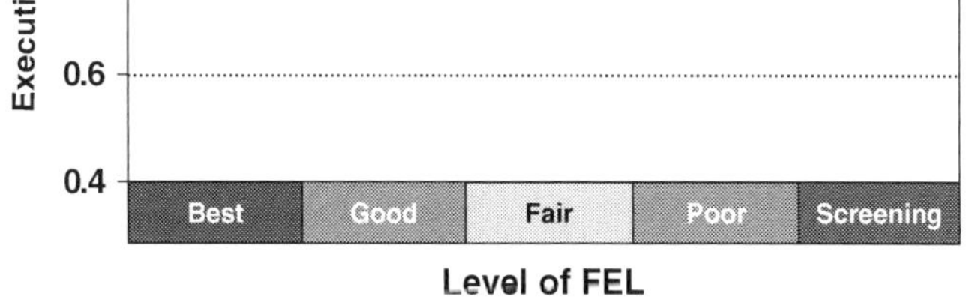

FIG. 9-27 Good FEL speeds execution time.

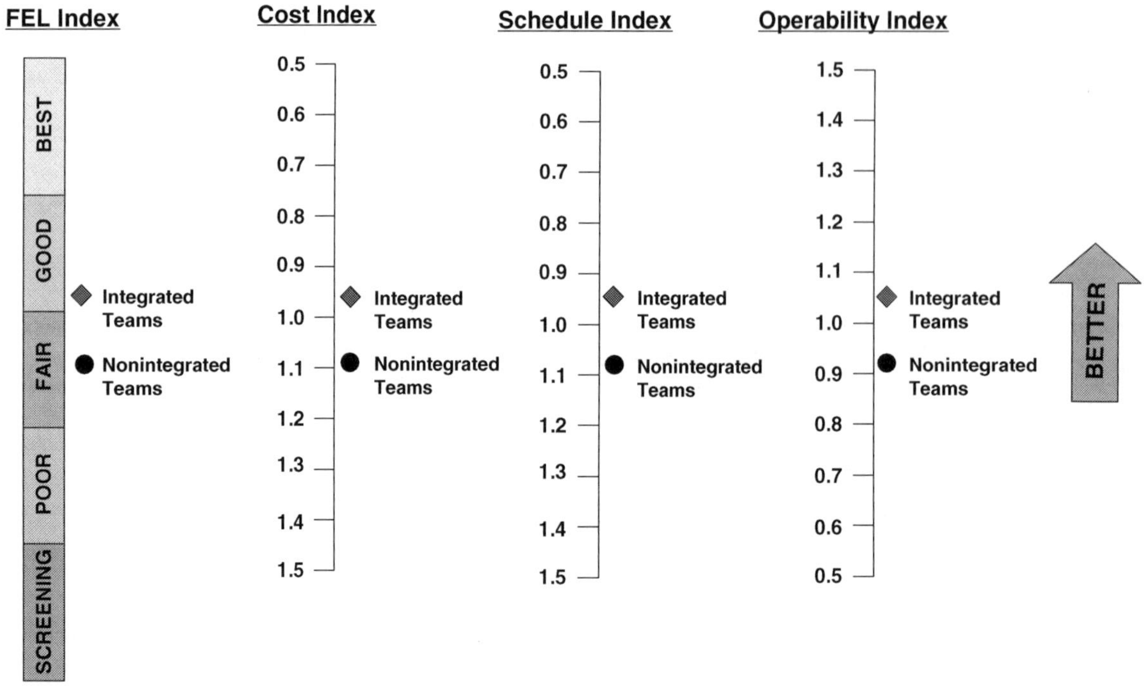

FIG. 9-28 Integrated teams result In better FEL and better overall performance.

Analysis, Inc., 32d Annual Engineering & Construction Contracting Conference, Sept. 28–29, 2000). These conclusions are depicted by Fig. 9-31. A major late change is defined by IPA's data to mean changes made after the start of detailed engineering and involving impacts greater than either 0.5 percent of the total project capital investment or 1 month in critical path project schedule.

These graphs illustrate why better project performance is produced through proactively seeking profit-improving changes as early as possible. One of the reasons for this observation is that operation, maintenance, and construction expertise is incorporated into the project at the very beginning—during FEL.

This means that the process design engineer should be working closely with these real-world experts as they design processes and their support systems. This also means that to improve overall project performance, achieving the best practical or highest level of definition during FEL is critical. Finally, this high level of definition results in a

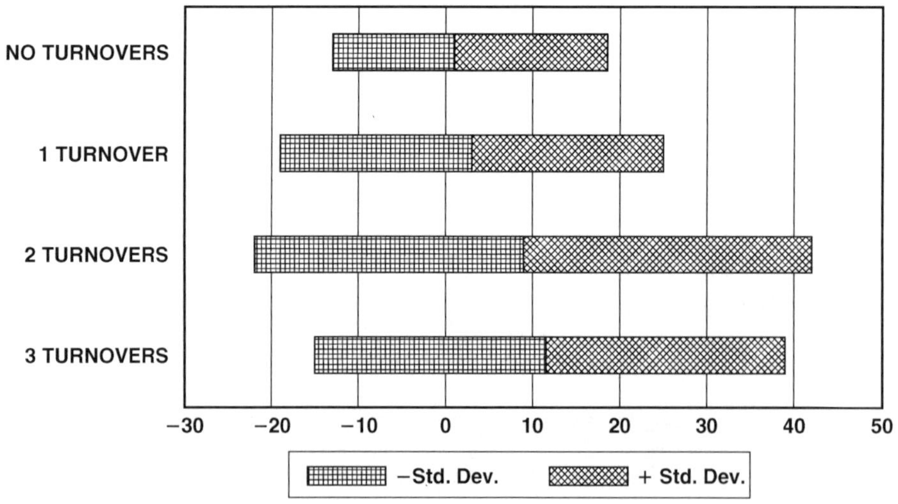

FIG. 9-29 Effects of project manager turnover on cost growth.

LEVEL OF FEL EFFORT	COST	SCHEDULE
High	−4%	−13%
Medium	−2%	+8%
Low	+16%	+26%
Difference	20%	39%

FIG. 9-30 Project performance versus level of FEL effort.

reduced number of changes during the EPC phase. These observations should be the critical goals of all project teams.

The size of capital project also has an influence on FEL outcome based on IPA statistics. IPA's data indicate that small projects benefit more from better project definition prior to the EPC phase than do major projects. The data also indicate that small projects typically have more late changes than do larger projects.

Figure 9-32 illustrates the effect of large projects versus small projects in terms of the impact of late changes (Merrow, E. W., Independent Project Analysis, Inc., 32d Annual Engineering & Construction Contracting Conference, Sept. 28–29, 2000). Figure 9-32 also illustrates that the level of FEL performance directly impacts the number of, and the consequences of, late changes made in projects of any size.

Typical FEL Deliverables Every process engineer assigned to a project should be acutely aware of which deliverables or end products are required by those who must commission their work. This should be very well understood by all parties prior to starting the work. Further, the splits of work (who will do which aspect of the work) must be well understood. Today, it is very common to have multiple operating companies form a joint venture to authorize major projects. It is also common for multiple engineering contractors to form joint ventures to execute the engineering for the FEL phases of the project.

Typical Conceptual Phase (FEL-1) Deliverables These are listed below. Each project will customize these deliverables to suit the particular needs of the project. There is no such thing as a "standard" FEL. Therefore, the process engineer must understand what the details are for each deliverable needed, what the *minimum* level is for the engineering required to meet those requirements, and in which formats that information and data will be needed.

- Strategic business assessment
- Key technology selected and risk identified
- Market assessment for feed, products, and capacity
- Potential sites identified and under evaluation
- Cost estimate (±40 percent)
- Preliminary project milestone schedule
- Block flow diagrams completed
- Process cases identified
- Critical long-lead equipment identified
- Value-improving practices reports

Typical Feasibility Phase (FEL-2) Deliverables These are listed below. In this phase, emphasis is on determining the best technical and economic flow scheme, as well as the support systems required to provide the necessary annual production rate at the sales quality required. The focus for the process engineer should be on confirming the number and type of process and technology studies needed, as well as the number of alternate cases required to be evaluated and/or simulated.

- Strategic business assessment
- Project schedule level 1
- Cost estimate (±25 percent)
- Overall project execution strategy
- Contracting and purchasing strategies
- Permitting and regulatory compliance plan
- Soil survey and report
- Project alternatives analysis
- Process flow diagrams for selected option(s)
- Preliminary utility flow diagrams and balances
- Preliminary equipment list and equipment load sheets
- Materials of construction
- Process hazards analysis report
- Value-improving practices reports

Typical Definition Phase (FEL-3) Deliverables These are listed below. In this phase, emphasis is typically on optimizing the best flow scheme and support systems combination. This optimum includes consideration of the plot plan and equipment arrangements for the entire facility. Process optimization cannot be done in isolation. Significant and continuous interaction with operations, maintenance, and construction experts always produces the best results. The emphasis in this phase is on achieving the best practical level of project definition and a good-quality project estimate of +/−10 percent. This level of project definition and cost estimate quality is normally required in order to present to management a candidate project which has the right combination of overall risk and projected economic performance, and thereby secure an AFE.

- Strategic business assessment
- Detailed EPC phase project execution plan
- Detailed EPC phase project master schedule

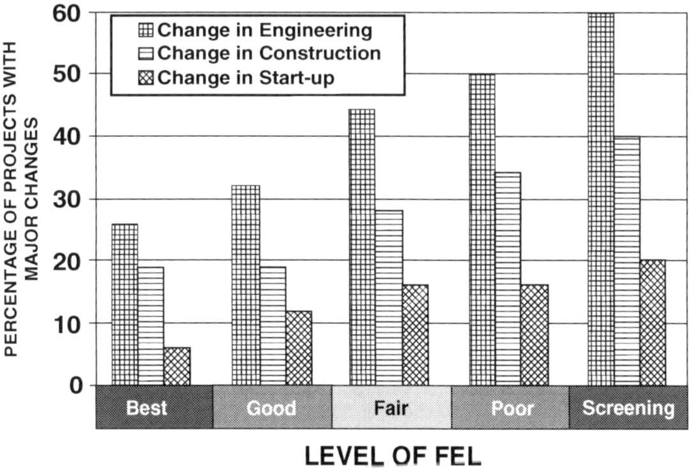

FIG. 9-31 Good FEL drives late changes down.

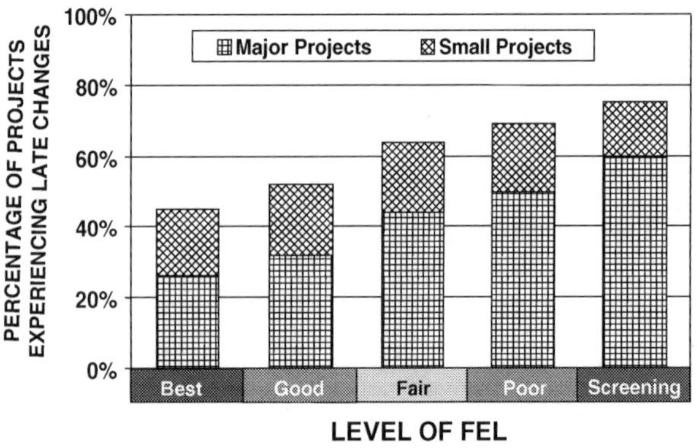

FIG. 9-32 Rate of late major changes is higher for small projects.

- Completed environment permit submittal
- Project plan/project execution plan
  - *a.* Cash flow plan for EPC phase
  - *b.* Training, commissioning, and start-up plans
  - *c.* Contracting plans
  - *d.* Materials management plan
  - *e.* Safety process and quality management plan
- Cost estimate (±10 percent)
- Finalized utility flow diagrams and balances
- P&ID's—issue IPL (issue for plant layout)
- Plot plans and critical equipment layouts
- Equipment list and equipment data sheets
- Single-line electrical diagrams
- Control system summary and control room layout
- Materials of construction
- VIPs reports
- Hazard and operability studies (HAZOP) report

VALUE-IMPROVING PRACTICES

GENERAL REFERENCES: Independent Project Analysis, Inc. (IPA), http://www.ipaglobal.com. The Construction Industry Institute (CII), University of Texas at Austin, http://construction-institute.org. Lavingia, Jr., N. J., Improve Profitability Through Effective Project Management and TCM, 36th Annual Engineering & Construction Contracting Conference, Sept. 4, 2003. KBR Value Improving Practices Program, 1995 through 2005, http://www.halliburton.com/kbr/index.jsp. PDRI: Project Definition Rating Index—Industrial Projects, Construction Industry Institute (CII), University of Texas at Austin, http://construction-institute.org, July 1996. Society of American Value Engineers International (SAVE), http://www.value-eng.org. KBR experience.

Introduction Value-improving practices (VIPs) are formal structured practices applied to capital projects to improve profitability (or "value") above that which is attained through the application of proven good engineering and project management practices. VIPs are formal analyses of project characteristics and features performed by small multidisciplinary teams at identified optimum times during the engineering design and development of capital projects.

Application of VIPs to capital projects has been statistically proved to significantly improve project profitability according to Independent Project Analysis, Inc. (IPA) and the Construction Industry Institute (CII). IPA data presented in Fig. 9-33 have been gathered from many capital projects since 1987. These data indicate that about 2.5 percent reduction in the relative capital cost can be expected for high-performing projects due to implementation of good front-end loading work processes. The best-performing projects are often referred to as "Best Practical" or "Best in Class" projects and represent the upper 20 percent of projects.

However, when FEL improvement is combined with rigorous application of VIPs, the project performance improves to about 10 percent in reduction of relative capital cost (Lavingia, Jr., N. J., Improve Profitability Through Effective Project Management and TCM). Experience of at least one major engineering contractor indicates that about a 20 percent capital cost improvement can be expected through judicious use of their modified VIPs (KBR Value Improving Practices Program, 1995 through 2005). These improved results come about from continual adaptation and improvement of the VIPs themselves to maintain their relevance and ability to improve projects above what the project teams can accomplish by themselves.

The VIPs that have been statistically verified by IPA benchmarking of capital projects are listed below. Each has a different purpose and focus, but all produce project profitability improvements that the project team cannot achieve on its own.

- Classes of facility quality
- Technology selection
- Process simplification
- Constructability
- Customization of standards and specifications
- Energy optimization
- Predictive maintenance
- Waste minimization
- Process reliability simulation
- Value engineering
- Design to capacity
- 3D-CAD

Selection of the most applicable VIPs to be performed during a specific project is the focus of the VIP planning session, which should be held just following project kickoff. Figure 9-34 presents the optimal times during a large project for consideration of VIPs. The duration for FEL has been assumed to be 12 months. Every project will have a

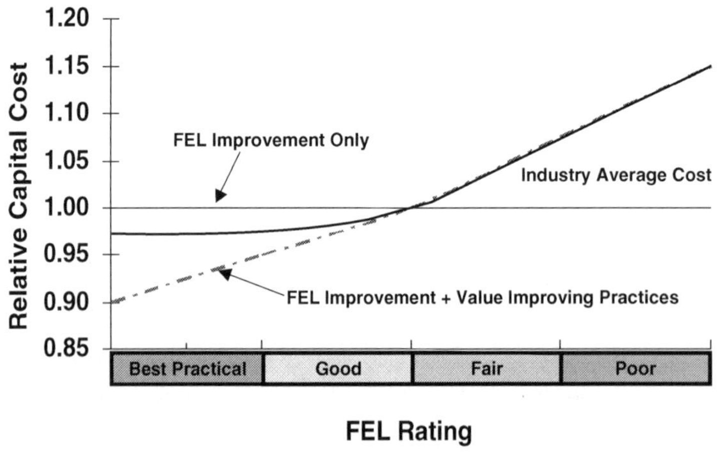

FIG. 9-33 FEL and VIPs drive lower capital investment.

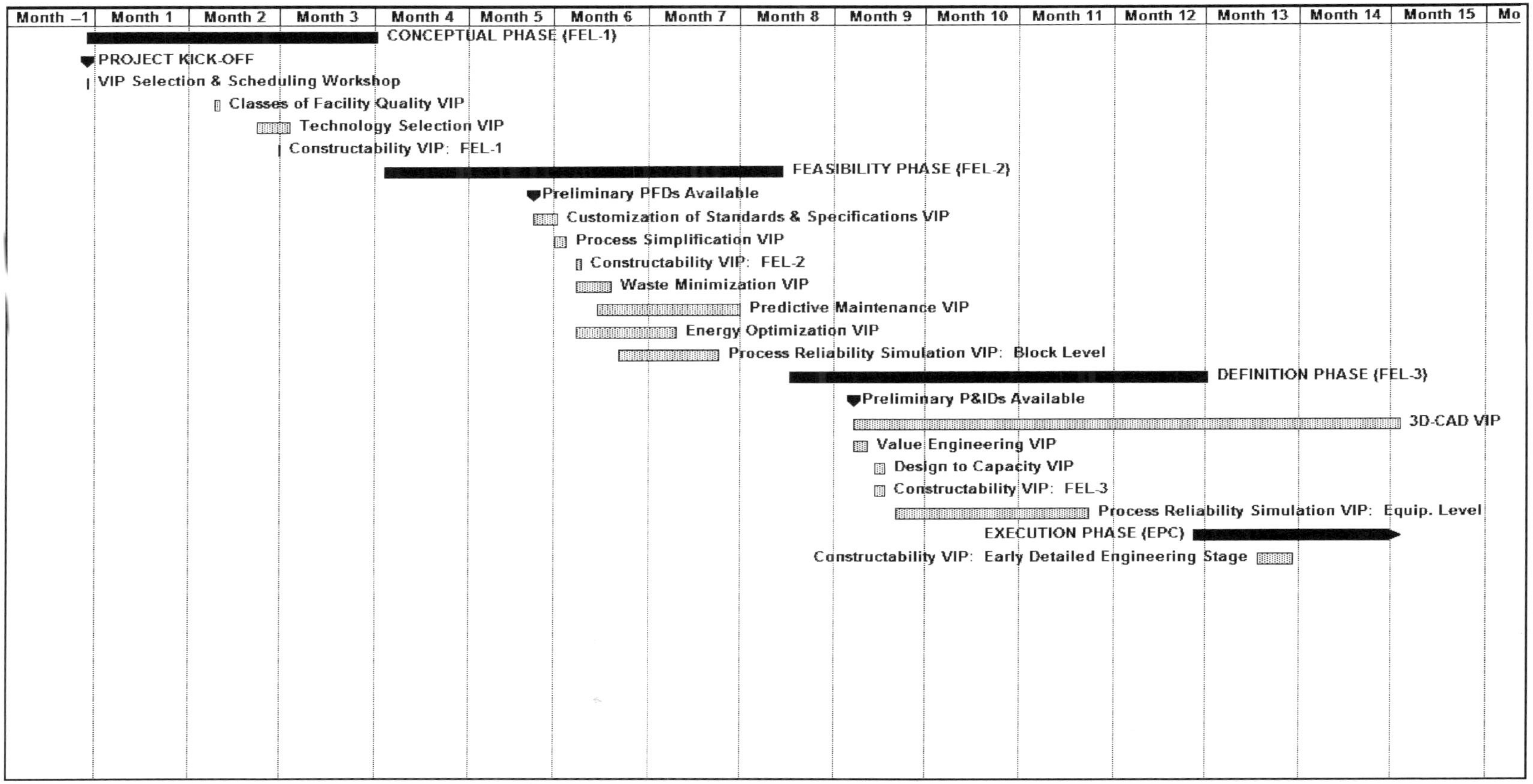

FIG. 9-34 Typical VIP implementation relationship.

unique duration for FEL with each phase dictated by the owner organization. The best times to conduct the VIP workshops should be considered at the project outset. Key anchor points for VIPs are the first appearance of the process flow diagrams (PFDs). The second anchor point for VIPs is the issue of preliminary piping and instrumentation diagrams (P&IDs).

When capital projects are benchmarked by third-party organizations such as IPA or through CII's Project Definition Rating Index (PDRI: Project Definition Rating Index—Industrial Projects, Construction Industry Institute), the implementation of applicable VIPs to the project is a part of their analysis. All the VIPs, when properly implemented, focus on producing better project definition and resultant economic improvements. The level of project definition achieved during FEL phases of the project is the focus of such benchmarking efforts.

VIPs as a group of practices are often described by their characteristics:

- Out-of-the-ordinary practices are used to improve cost, schedule, and/or reliability of capital projects.
- They are used primarily during FEL project phases.
- Formal and documented practices involve repeatable work processes.
- All involve a formal facilitated workshop to confirm the value gained to the project and to formally approve VIP team recommendations.
- All involve stated explicit support from the owner's corporate executive team.
- VIPs must be performed by a trained experienced VIP facilitator—someone who is not a member of the project team
- VIPs are characterized by statistical links between the use of the practice and better project performance which are demonstrated, systematic, repeatable, and proven correlations

VIPs are also further clarified by what they are not, as below.

- Just "good engineering"
- Simple brainstorming or strategy sessions
- Business as usual
- A special look at some aspect of the project
- Cost reduction or scope reduction exercises
- PFD or P&ID reviews
- Safety reviews
- Audits
- Project readiness reviews

VIP Descriptions

Classes of Facility Quality VIP The class of facility quality VIP determines the appropriate classes of facility quality that would produce the highest value or profitability in terms of

- Capital investment (CAPEX)
- Planned facility life
- Expandability
- Level of automation
- Equipment selection,
- Operating expense (OPEX)
- Environmental controls
- Capacity
- Technology

This VIP individually confirms the best overall design philosophy for the project team, for each of the parameters listed above. Here, the designer first learns how aggressive the owner organization wants the facility design and operation to be in terms of overall risk. For example, if the plant is to have the lowest possible OPEX, then the designer will incorporate greater levels of automation, instrumentation, and robustness of mechanical design in the overall facility.

The results of this VIP are used by the project management team to update its project execution plan for each FEL phase. The class of facility quality VIP provides the best results when conducted prior to executing any other VIP effort in the conceptual phase (FEL-1) of the project.

Technology Selection VIP The technology selection VIP is the application of evaluation criteria aligned with the project's business objectives to identify manufacturing and processing technology that may be superior to that currently used. The goal is to ensure that the technology suite finally selected is the most competitive available. This requires a systematic search, both inside and outside the operating company's organization, to identify emerging technology alternatives. This formal facilitated process is also meant to ensure due diligence for all parties involved and that all emerging and near-commercial alternative technologies for accomplishing a particular processing function are objectively considered. This VIP is most commonly applied at the unit operation level, although it has also been successfully applied down to the major equipment level (KBR Value Improving Practices Program, 2000 through 2005). This VIP is particularly effective for combating the NIH syndrome ("not-invented-here").

The goals of this VIP are to document which technology evaluation criteria are applicable and then to conduct a formal technology screening and evaluation assessment. The result is a prioritized listing of technology options for each selected application for the project. The preferred time to execute this VIP is the midpoint in the conceptual phase (FEL-1).

Process Simplification VIP The process simplification VIP uses the value methodology and is a formal, rigorous process to search for opportunities to eliminate or combine process and utility system steps or equipment, ultimately resulting in the reduction of investment and operating costs. The focus is the reduction of installed costs and critical path schedule while balancing these value improvements with expected facility operability, flexibility, and overall life cycle costs.

The process simplification VIP does far more than just evaluate and simplify processing steps. This very productive VIP ensures that low- or zero-value functions or equipment included in the project scope are challenged by experienced world-class experts and eliminated, if possible. This VIP tries to systematically differentiate "wants" from "needs" and remove the "wants." It can be especially effective for providing a neutral professional environment for identifying and challenging "sacred cows" and then removing them. Removal of these low- or zero-value functions yields significant profitability improvements to the overall project. Process simplification results in

- Reduced capital costs (CAPEX)
- Improved critical path schedule
- Reduced process inventory
- Increased yields
- Reduced operating and maintenance costs (OPEX)
- Increased productivity
- Incremental capacity gains
- Reduced utility and support systems requirements
- Reduced waste generation

Process simplification is executed in a formal workshop with a trained experienced facilitator. This VIP should always include key participants from each of the project owner's organizations, the engineering contractor organization, key third-party technology licensors, and equipment or systems vendors, where possible. One or more "cold eyes" reviewers or subject matter experts, who have extensive experience, should be included to provide an objective and unbiased perspective.

This VIP also provides a means for integrating overall plantwide systems. The process simplification VIP is typically performed during the feasibility phase (FEL-2) after the preliminary PFDs and heat and material balances become available. However, for very large and complex projects, considerable value has been gained by also performing this VIP at the midpoint or later in the conceptual phase (FEL-1).

Constructability VIP The constructability VIP is the systematic implementation of the latest engineering, procurement and construction concepts and lessons learned, which are consistent with the facility's operations and maintenance requirements. The goal is to enhance construction safety, scope, cost, schedule, and quality.

Since the constructability VIP has seen widespread implementation in industry for capital projects over the last 20 years, in order for this VIP to remain consistent with the definition of a VIP (i.e., above what project teams can do on their own), at least one large engineering and construction company has enhanced this VIP to include a formal facilitated workshop that seeks profitability improvements above those already identified by the project team in the course of its normal work (KBR Value Improving Practices Program, 2001 through 2005). Both work processes described below are mutually additive, flexible, and compatible.

The traditional constructability work process includes the following characteristics:

- Starts at the FEL-1 phase and continues through facility start-up
- Is an ongoing structured program

- Optimizes the combined use of operations, maintenance, engineering, procurement, key vendors, and construction knowledge and experience
- Enhances the achievement of project objectives
- Has construction experts working with the engineering and procurement process that results in construction safety, cost, schedule, and quality savings
- Uses on-project and off-project expertise

The enhanced constructability VIP adds the following to the traditional approach:

- Includes a formal facilitated workshop
- Is held in every engineering phase of the project with a focus on the pertinent aspects of that phase
- Identifies value improvements and their benefits above those already being considered by the traditional constructability work process
- Focuses on the systematic implementation of the latest engineering, procurement, construction concepts, and lessons learned
- Involves a detail review of planning, design, procurement, fabrication, and installation functions to achieve the best overall project safety performance, lowest CAPEX, and the shortest reasonable schedule
- Applies operations and maintenance requirements and expertise
- Includes considerations for operability and maintainability
- Enhances construction safety, scope, cost, schedule, and quality

A formal constructability VIP workshop conducted in the conceptual phase (FEL-1) should focus on the overall project construction strategies regarding site layout, construction and turnaround laydown areas, access to the site for large equipment and modules, modularization, sequencing of heavy lifts, limitations regarding procurement, limitations regarding fabrication and transport, area labor limitations, and coordination with any existing or nearby structures or facilities.

A formal constructability VIP workshop conducted in the feasibility phase (FEL-2) should focus on more specific topics of layout optimization, using a preliminary plot plan and equipment layout for the project. Considerations should include optimum site layout in terms of construction laydown areas; optimum equipment arrangement to reduce piping and steel for structures and piperack; specific sizes and weights for modules; which components will be included in each module; crane locations for heavy lifts; equipment requiring early purchase to allow project schedule to be achieved; further analysis of limitations regarding procurement, fabrication, and area labor availability; and precommissioning, commissioning, and start-up considerations.

A formal constructability VIP workshop conducted in the definition phase (FEL-3) focuses on even greater detail for what was discussed above. In the detailed engineering stage (EPC phase), considerable detail will be reviewed to evaluate how the project can best construct what will be needed. Here, significant application of detailed lessons learned is reviewed and considered.

Constructability VIP workshops should be formal facilitated workshops drawing on personnel from operations, maintenance, and construction in addition to project and owner organization representation.

Customization of Standards and Specifications VIP The customization of standards and specifications VIP is a direct and systematic method to improve project value by selecting the most appropriate codes, standards, and specifications for the project. The goal is to make helpful changes to meet the actual project requirements, ensuring that the codes, standards, and specifications selected do not exceed those required for the project, and maximizing the use of specifications from equipment vendors to obtain the best overall value. This VIP is beyond typical good engineering practices and should not be confused with ongoing systematic improvements in corporate standards and specifications, or with required identification of applicable procurement specifications to be used for the project. This formal VIP takes a combination of project owner and engineering contractor corporate specifications and aggressively seeks profitability improvements consistent with the project's goals and limitations. This VIP maximizes the procurement of off-the-shelf equipment over equipment customized for the project.

Industry experience indicates that project-specific "Fit for Purpose" standards and specifications on the average cost less than the general application of traditional standards. This VIP is best performed early in the feasibility phase (FEL-2), and should include project team members involved from both the project owner and engineering contractor, as well as appropriate suppliers of major packaged subsystems, modularized equipment, etc.

Energy Optimization VIP The energy optimization VIP is the systematic process for the evaluation of the thermal efficiency of a process (or multiple subunits within a larger process or facility). The goal is to improve the economic utilization of energy. This optimization starts by using the "pinch" technology branch of process energy integration (energy pinch) to identify better process energy exchange options.

Energy pinch (usually just called *pinch*) is a methodology for the conceptual design of process heating, utility, and power systems. Pinch allows the maximization of energy utilization within a process, while minimizing the use of plant utilities. Such minimization is achieved by reusing energy, via process stream–to–process stream heat exchange. A pinch analysis is performed by analyzing the tradeoff between the energy which can be recovered and the additional capital costs which must be added to do so. It includes the project's design and thermodynamic constraints (performance targets) available during this preliminary design phase.

The benefits of pinch technology include lower operating costs, occasionally reduced capital cost, improved operability/flexibility, increased throughput, and site-specific process optimization and reduced emissions. Pinch technology can be applied for both grass-roots and retrofit applications. Typical applications include process heat integration as well as sitewide heat and power integration. However, this methodology is profitably applied to the optimization of high-value complex mass flow problems, such as refinery hydrogen network optimization (hydrogen pinch) and wastewater minimization (water pinch).

Once the minimum theoretical energy requirements and applicable process options have been determined, a formal facilitated workshop follows to modify the process or facility to bring the design closer to the thermodynamic optimum within project economic constraints.

The energy optimization VIP is most beneficial for processes where energy and related capital expense are a relatively large fraction of the total operating cost. The benefits result in reduced energy requirements and environmental emissions in balance with project economics. This VIP should be implemented in the feasibility phase (FEL-2) when preliminary PFDs and heat and material balances are available.

Predictive Maintenance VIP The predictive maintenance VIP is the proactive use of sensors and associated controls to monitor the machinery mechanical "health," using both current state and historical trends, to optimize effective planning of all shutdowns and maintenance, thereby detecting equipment abnormalities and diagnosing potential problems before they cause permanent equipment damage. Examples include real-time corrosion monitoring and equipment vibration monitoring. This additional instrumentation is generally economically justified in the case of critical equipment items and key operations. Predictive maintenance reduces maintenance costs, improves the confidence of extending time between turnarounds, improves reliability, and provides a more predictable maintenance schedule for key process equipment. It also minimizes the amount of remaining equipment life that is lost through using only preventive maintenance practices. Preventive maintenance is an older practice which is limited to periodic inspections and repairs to avoid unplanned equipment breakdowns.

For the predictive maintenance VIP to be effective, maintenance personnel from the project owner's organization must be involved in determining key predictive maintenance requirements. Suppliers of critical equipment items (i.e., compressors) are also important participants in this process.

The predictive maintenance VIP is considered by some operating companies and engineering contractors to have become standard practice. For those where it is not already standard practice, this VIP should be initiated in the feasibility phase (FEL-2) and concluded with a formal facilitated workshop and report of recommendations to the project management team.

Waste Minimization VIP The waste minimization VIP involves a formal process stream-by-stream analysis to identify ways to eliminate or reduce the generation of wastes or nonuseful streams within the chemical process itself. For those streams not eliminated or converted to salable by-products, it provides the method for managing the resulting wastes. This VIP incorporates environmental requirements into the facility design and combines life cycle environmental benefits and positive economic returns through energy reductions, reduced end-of-pipe treatment requirements, and improved raw material yields. The waste reduction hierarchy is to

- Eliminate or minimize the generation of waste through source reduction
- Recycle by use, reuse, or reclamation those potential waste materials that cannot be eliminated or minimized
- Treat all waste that is nevertheless generated to reduce volume, toxicity, or mobility prior to storage or disposal

This VIP is considered by some engineering contractors to have become standard practice. For those where it is not standard practice, the waste minimization VIP should be executed in a formal workshop with an experienced facilitator with project owner and engineering contractor representatives always involved. A "cold eyes" reviewer with extensive experience should also be included to add a nonbiased perspective. The waste minimization VIP should be implemented at the feasibility phase (FEL-2) when preliminary PFDs and heat and material balances are available.

Process Reliability Simulation VIP The process reliability simulation VIP is the use of reliability, availability, and maintainability (RAM) computer simulation modeling of the process and the mechanical reliability of the facility. A principal goal is to optimize the engineering design in terms of life cycle cost, thereby maximizing the project's potential profitability. The objective is to determine the optimum relationships between maximum production rates and design and operational factors. Process reliability simulation is also applied for safety purposes, since it considers the consequences of specific equipment failures and failure modes.

This VIP is typically led by an engineer experienced in plant operations and the use of the RAM simulation modeling software. The VIP should also directly involve the project owner since that organization would most often supply the historical operating and maintenance information required for the development of the simulation model. This process provides the project team with a more effective means of assessing, early in the design, the cost/benefit impact of changes in design, identification of bottlenecks in the system, simulation of key operating scenarios, determination of equipment-sparing needs, training and maintenance requirements of a facility.

The process reliability simulation VIP should be initiated in the feasibility phase (FEL-2) to produce a block-level RAM model. Based on the results of that model, a more detailed equipment-level RAM model should be developed starting in the definition phase (FEL-3).

Value Engineering VIP The value engineering VIP is a flexible, organized, multidisciplinary team effort directed at analyzing the functions, issues, and essential characteristics of a project, process, technology, or system. The goal is to satisfy those functions, issues, and essential characteristics at the lowest life cycle cost. The value engineering VIP rigorously examines what is needed to meet the business objectives of a project and the elimination of non-value-adding investment. An open-minded attitude by participants is required to effectively remove unneeded scope and in doing so reduce the installed costs of the project. This VIP tries to systematically differentiate "wants" from "needs" and remove the "wants." Tests for non-income-producing investments include redundancy, overdesign, manufacturing add-ons, upgraded materials of construction, and customized design versus vendor standards.

The value engineering VIP also ensures that low- or zero-value functions or equipment included in the project scope are challenged to be the highest value possible for the project. Removal of these low- or zero-value functions from the project scope, if possible, will most likely yield significant profitability improvements to the overall project. These can encompass the following:

- Misalignment of unit or system capacity or operations capability with respect to the overall facility
- Overly conservative assumptions of the basic design data
- Overly conservative interpretation of how the facilities will be used during peak, seasonal, or upset conditions
- Preinvestment included in the project scope that may not be value added
- Overdesign of equipment or systems to provide uneconomic added flexibility

The value engineering VIP is executed in a formal workshop with a trained experienced technical workshop facilitator. Both the project owner and the engineering contractor are always involved. Third-party licensors and equipment/system vendors should be included where applicable. One (or more) cold-eyes reviewer with extensive experience should also be included to provide an unbiased perspective.

This VIP leverages the growing accumulation of more detailed project knowledge to test the value of earlier, more generalized scope assumptions. It also tests the presumed added value of different stakeholder requirements, which have influenced the evolution of the project scope. This highly adaptable VIP results in reduced capital costs (CAPEX), improved critical path schedule, reduced process inventory, increased yields, reduced operating and maintenance costs (OPEX), increased productivity, incremental capacity gains, reduced utility and support systems requirements, and reduced waste generation.

The value engineering VIP should be conducted in the definition phase (FEL-3) when the first issue of P&IDs is available.

Design to Capacity VIP The design to capacity VIP systematically evaluates the maximum capacity of major equipment, ancillary piping, valving, instrumentation, and associated engineering calculations and guidelines. The goal is to improve life cycle costs (profitability) by eliminating preinvestment and overdesign. This VIP requires the systematic and formal evaluation of the maximum capacity of each piece of equipment instead of the traditional practice of designing with an extra safety factor or margin to allow for additional catch-up capacity or some future production increase. The goal is also to eliminate overdesign in both calculations and engineering guidelines. This VIP is conducted as a facilitated workshop with both project owner and engineering contractor representation.

This VIP reduces capital investment by confirming minimum required capacities and flexibility necessary only to meet current project business objectives. The workshop drills down to each specific system and subsystem and finally scrutinizes the design of each equipment item. This workshop is often combined with the value engineering VIP, which overlaps significantly. The design to capacity VIP should be conducted in the definition phase (FEL-3) when the first issue of P&IDs is available.

3D-CAD VIP The 3D-CAD VIP is the creation of a detailed three-dimensional (3D) computer model depicting the proposed process and associated equipment along with the optimized plant layout and specific equipment arrangements and orientations. The 3D model can then be used to generate computerized interference checks of bulk material configurations and equipment and extraction of error-free fabrication drawings and material quantities. The goals of this VIP are to reduce engineering and construction rework, improve operability and maintainability, and confirm the incorporation into the design of advantageous human factors (a.k.a. ergonomics) focused on ease of operation and maintenance.

A number of industry-accepted state-of-the-art 3D computer-aided design (CAD) systems have been used for this purpose. The specification-driven 3D-CAD system allows a computer model to be built to allow extraction of drawings from the model for fabrication and erection. The extracted drawings are enhanced in their accuracy by the computer interference detection system which greatly reduces field rework.

The principal benefit of utilizing 3D-CAD is the ability to produce an electronic model that accurately resembles the completed facility. This enables project teams, clients, and constructors to review and agree on the plant design before construction starts. The model can then be used to generate interference checks of bulk material configurations and equipment, as well as the extraction of error-free fabrication drawings and material quantities. The system also utilizes its design review capabilities to confirm proposed designs and obtain approvals from key project stakeholders and owners.

This VIP is considered by most major engineering contractors to have become standard practice. The 3D-CAD VIP model development should be initiated in the feasibility phase (FEL-2) after the plot plan has been finalized and the first issue of P&IDs is available.

VIP Planning and Implementation Each VIP has a unique character, and it should be performed at a certain time and in a certain way to produce the best results for the project. Part of the power of VIPs is that they can be used to improve the overall economics of the project without the need for inordinate additional time or expense. Ironically, the return on investment (ROI) for the cost of implementing each VIP is usually much greater than that ROI for the overall proposed project. For one engineering contractor, *the typical ROI for* *implementing VIPs is at least one order of magnitude higher* (KBR Value Improving Practices Program, 2000 through 2005).

It is important to reiterate that the benefits of VIPs cannot be realized by just executing "good engineering." The application and implementation of VIPs to any project must have the explicit commitment of the owner's corporation executives. VIP execution must be deliberately and carefully planned in the initial phase of the project. For all projects, this VIP planning meeting should take place immediately following project kickoff. The project management team and the selected VIP facilitator should

- Confirm which VIPs should be applied to the project and when
- Incorporate the planned VIPs into the project scope of work and schedule
- Determine the required workshop resources and best combination of engineering, operations, maintenance, construction, and other expertise for each selected VIP workshop team

VIPs That Apply the Value Methodology Nearly all VIPs are conducted only once in a project at a "sweet spot" where maximum benefit is found. For example, the process simplification VIP is anchored at the first appearance of the preliminary PFDs, while the value engineering VIP and the design to capacity VIP are anchored at the first issue of the P&IDs. Both of these apply the value methodology [Society of American Value Engineers International (SAVE)] that has produced excellent results in industry for more than 55 years. The typical approach and steps for these three unique VIPs are presented below.

Preparation and Planning Before the VIP is begun, the goals, objectives, and scheduled time for the formal workshop must be agreed upon by the integrated project management team. The workshop facilitator must ensure that all the information required for the workshop is available and that the workshop team members have been fully briefed on the VIP's objectives, methodology, and expectations.

The Formal Workshop The formal workshop is always structured to make maximum use of the multi-disciplinary team's time and effort. Such workshops typically require no less than 2 days and as many as 5 days depending on the size and complexity of the project. The required workshop length should be determined by the VIP facilitator. A typical process simplification VIP, value engineering VIP, and design to capacity VIP workshop includes the following phases of a typical "job plan" that are supported by the Society of American Value Engineers International.

The information phase In this phase, team members review important background materials and confirm their understanding of the basis for the design of the project, the constraints, and the sensitivity of the relevant capital and operating costs. Here, incorporating important unresolved project issues into the workshop produces more meaningful financial and technical results. Discussion of the issues' validity and basis are determined during the first day of the workshop. These issues very often become some of the best brainstorming targets for cost and schedule reduction ideas.

A very specific and structured methodology is used which is known as the *Function Analysis System Technique (FAST)*. This function analysis diagramming illustrates the logical or functional relationships and dependencies between different process systems and project activities. These diagrams are then reviewed and critiqued together with the associated costs of selected groups of process functions or project functions. The function analysis can be performed at this stage, but often time can be saved by preparing a draft of these FAST diagrams prior to the workshop with a small group of the workshop team members.

The speculation phase Once the pertinent information and issues have been reviewed and the important functions of each process and project step identified, the team is encouraged to speculate on alternative methods to perform each function and to solve each major project issue. Brainstorming sessions within a creative environment encourage the team to strive for new and innovative ideas.

The conceptual phase The team then reviews the ideas against relevant project criteria such as potential impact on long-term project economics, impact on operations and maintenance costs, effect on the capital cost for the project, validity to the project scope of work, technical risks associated with implementation of the new concept, impact on project schedule, and cost required to implement the improvement. Each study has specific criteria against which proposed alternatives are judged. The ideas are weighted, sorted, grouped, linked, and ranked so that the best of the technically viable ideas are efficiently identified for further detailed study.

The feasibility phase The ideas with the most merit are developed into preliminary two-page written proposals with potential benefits approximated as part of the workshop. Performing this important activity following the formal workshop has been shown to often result in significant loss of potential for implementation. The VIP team expands the ideas ranked highest to obtain additional technical and economic insights and information to support the idea. The proposals are then presented internally, to the assembled VIP workshop team, and discussed to determine whether the ideas retain sufficient technical and economic merit to be recommended by the VIP team to a separate steering committee or project management team.

Experience indicates that having the VIP team perform this stage within the formal workshop produces the best results.

The presentation phase The VIP team formally presents the profitability recommendations consistent with the objectives and constraints of the workshop and their implementation plans to the steering committee or the project management team. The steering committee then approves those recommendations that pass muster and authorizes the project team to begin the implementation effort. Often, this approval is conditional on early validation by subject matter experts within the project owner's organization, but not present within the workshop. This external feasibility check is meant to provide support to the project team for any additional resources and schedule time needed to fully incorporate the improvements into the project scope of work.

Report and follow-up After completing the intensive VIP workshop, the workshop facilitator completes the written VIP final report for the project record. During this time, the project management team assigns each approved recommendation to a member of the project team, estimates the engineering time and resources required to incorporate the improvement into the project scope of work, and communicates the results of the VIP within the integrated project team. This follow-up action plan creates a very positive and cost-conscious attitude within the project team that leads to further improvements in project value.

Sources of Expertise VIP workshops should be planned and led by a trained experienced facilitator who has significant experience in effectively conducting such VIP workshops. Technical expertise for VIP workshops should be a combination of senior project team members and subject matter experts from the operating company's organization, the engineering contractor's organization, licensed technology providers, and any key fabrication or installation subcontractors to be used. Figure 9-35 illustrates the best balance of expertise for VIP workshops (KBR Value Improving Practices Program, 1995 through 2005).

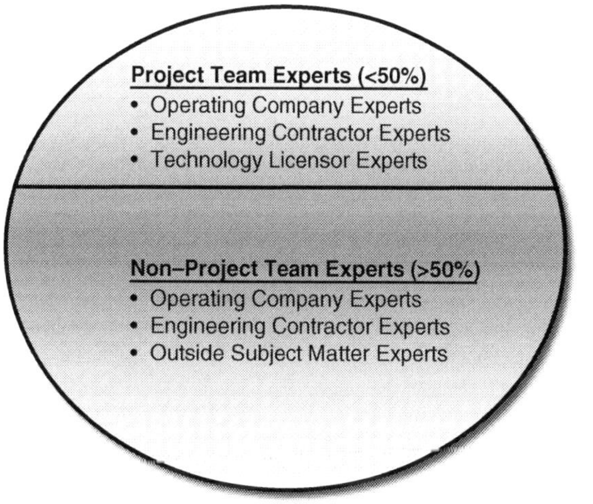

FIG. 9-35 The ideal VIP team makeup.

GLOSSARY

Accounts payable The value of purchased goods and services that are being used but have not been paid.

Accounts receivable Credit extended to customers, usually on a 30-day basis. Cash is set aside to take care of the probability that some customers may not pay their bills.

Accrual basis The accounting method that recognizes revenues and disbursement of funds by receipt of bills or orders and not by cash flow, distinguished from *cash basis*.

Administrative expense An overhead expense due to the general direction of a company beyond the plant level. It includes administrative and office salaries, rent, auditing, accounting, legal, central purchasing and engineering, etc., expenses.

Allocation of expenses A procedure whereby overhead expenses and other indirect charges are assigned back to processing units or to products on what is expected to be an equitable basis. All allocations are somewhat arbitrary.

Amortization Often used interchangeably with *depreciation*, but there is a slight difference depending on whether the life of an asset is known. If the period of time is known to be usually more than a year, this annual expense is *amortization*; however, if the life is estimated, then it is *depreciation*.

Annual net sales Pounds of product sold times the net selling price. *Net* means that any allowances have been subtracted from the gross selling price.

Annual report Management's report to the stockholders and other interested parties at the end of a year of operation showing the status of the company, its activities, funds, income, profits, expenses, and other information.

Appurtenances The auxiliaries to either process or nonprocess equipment: piping, electrical, insulation, instrumentation, etc.

Assets The list of money on hand, marketable securities, monies due, investments, plants, properties, intellectual property, inventory, etc., at cost or market value, whichever is smaller. The assets are what a company (or person) owns.

Balance sheet This is an accounting, historical tabulation of assets, liabilities, and stockholders' equity for a company. The assets must equal the liabilities plus the stockholders' equity.

Battery limit A geographic boundary defining the coverage of a specific project. Usually it takes in the manufacturing area of a proposed plant, including all process equipment but excluding provision for storage, site preparation, utilities, administrative buildings, or auxiliary facilities.

Bonds When one purchases a bond, the company (or person) acquires an interest in debt and becomes a creditor of the company. The purchaser receives the right to receive regular interest payments and the subsequent repayment of the principal.

Book value Current investment value on the company books as the original installed cost less depreciation accruals.

Book value of common stock Net worth of a firm divided by the number of shares of common stock issued at the time of a report.

Break-even chart An economic production chart depicting total revenue and total expenses as functions of operation of a processing facility.

Break-even point The percentage of capacity at which income equals all fixed and variable expenses at that level of operation.

By-product A product made as a consequence of the production of a main product. The by-product may have a market value or a value as a raw material.

Capacity The estimated maximum level of production on a sustained basis.

Capital ratio Ratio of capital investment to sales dollars; the reciprocal of capital turnover.

Capital recovery The process by which original investment in a project is recovered over its life.

Capital turnover The ratio of sales dollars to capital investment; the reciprocal of capital ratio.

Cash Money that is on hand to pay for operating expenses, e.g., wages, salaries, raw materials, supplies, etc., to maintain a liquid financial position.

Cash basis The accounting basis whereby revenue and expense are recorded when cash is received and paid, distinguished from accrual basis.

Cash flow Net income after taxes plus depreciation (and depletion) flowing into the company treasury.

Code of accounts A system in which items of expense or fixed capital such as equipment and material are identified with numerical figures to facilitate accounting and cost control.

Common stock Money paid into a corporation for the purchase of shares of common stock that becomes the permanent capital of the firm. Common stockholders have the right to transfer ownership and may sell the stock to individuals or firms. Common stockholders have the right to vote at annual meetings on company business or may do so by proxy.

Compound interest The interest charges under the condition that interest is charged on previous interest plus principal.

Contingencies An allowance for unforeseeable elements of cost in fixed investment estimates that previous experience has shown to exist.

Continuous compounding A mathematical procedure for evaluating compound interest based upon continuous interest function rather than discrete interest periods.

Conversion expense The expense of converting raw materials to finished product.

Corporation In 1819, defined by Chief Justice Marshall of the Supreme Court as "an artificial being, invisible, intangible and existing only in contemplation of law." It exists by the grace of a state, and the laws of a state govern the procedure for its formation.

Cost of capital The cost of borrowing money from all sources, namely, loans, bonds, and preferred and common stock. It is expressed as an interest rate.

Cost center For accounting purposes, a grouping of equipment and facilities comprising a product manufacturing system.

Cost of sales The sum of the fixed and variable (direct and indirect) expenses for manufacturing a product and delivering it to a customer.

Decision or decision making A program of action undertaken as a result of (1) an established policy or (2) an analysis of variables that can be altered to influence a final result.

Depletion A provision in the tax regulations that allows a business to charge as current expense a noncash expense representing the portion of limited natural resources consumed in the conduct of business.

Depreciation A reasonable allowance by the Internal Revenue Service for the exhaustion, wear and tear, and normal obsolescence of equipment used in a trade or business. The property must have a useful life of more than 1 year. Depreciation is a noncash expense deductible from income for tax purposes.

Design to cost A management technique to achieve system designs that meet cost parameters. Cost as a design parameter is considered on a continuous basis as part of a system's development and production processes.

Direct expense An expense directly associated with the production of a product such as utilities, labor, and maintenance.

Direct labor expense The expense of labor involved in the manufacture of a product or in the production of a service.

Direct material expense The expense associated with materials consumed in the manufacture of a product or the production of a service.

Distribution expense Expense including advertising, preparation of samples, travel, entertainment, freight, warehousing, etc., to distribute a sample or product.

Dollar volume Dollar worth of a product manufactured per unit of time.

Earnings The difference between income and operating expenses.

Economic life The period of commercial use of a product or facility. It may be limited by obsolescence, physical life of equipment, or changing economic conditions.

Economic value added The period dollar profit above the cost of capital. It is a means to measure an organization's value and a way to determine how management's decisions contribute to the value of a company.

Effective interest The true value of interest computed by equations for the compound interest rate for a period of 1 year.

Equity The owner's actual capital held by a company for its operations.

Escalation A provision in actual or estimated cost for an increase in equipment cost, material, labor, expenses, etc., over those specified in an original estimate or contract due to inflation.

External funds Capital obtained by selling stocks or bonds or by borrowing.

FEL (front-end loading) The process by which a company develops a detailed definition of the scope of a capital project that meets corporate business objectives.

FIFO (first in, first out) The valuation of raw material and supplies inventory, meaning first into the company or process is the first used or out.

Financial expense The charges for use of borrowed funds.

Fixed assets The real or material facilities that represent part of the capital in an economic venture.

Fixed capital Item including the equipment and buildings.

Fixed expense An expense that is independent of the rate of output, e.g., depreciation and plant indirect expenses.

Fringe benefits Employee welfare benefits; expenses of employment over and above compensation for actual time worked, such as holidays, vacations, sick leave, insurance.

Full cost accounting Method of pricing goods and services to reflect their true costs, including production, use recycling, and disposal.

Future worth The expected value of capital in the future according to some predetermined method of computation.

Goods manufactured, cost of Total expense (direct and indirect expenses) including overhead charges.

Goods-in-process inventory The holdup of product in a partially finished state.

Goods sold, cost of The total of all expenses before income taxes that is deducted from income (revenue).

Grass-roots plant A complete plant erected on new site including land, site preparation, battery-limits facilities, and auxiliary facilities.

Gross domestic product An indicator of a country's economic activity. It is the sum of all goods and services produced by a nation within its borders.

Gross margin (profit) Total revenue minus cost of goods manufactured.

Gross national product An economic indicator of a country's economic activity. It is the sum of all the goods and services produced by a nation both within and outside its borders.

Income Profit before income taxes or gross income from sales before deduction of expenses.

Income statement The statement of earnings of a firm as approximated by accounting practices, usually covering a 1-year period.

Income tax The tax imposed on corporate profits by the federal and/or state governments.

Indirect expenses Part of the manufacturing expense of a product not directly related to the amount of product manufactured, e.g., depreciation, local taxes, and insurance.

Internal funds Capital available from depreciation and accumulated retained earnings.

Inventory The quantity of raw materials and/or supplies held in a process or in storage.

Last in, first out (LIFO) The valuation of raw materials and supplies, meaning the last material into a process or storage is the first used or out.

Leverage The influence of debt on the earning rate of a company.

Liabilities An accounting term for capital owed by a company.

Life cycle cost Cost of development, acquisition, support, and disposal of a system over its full life.

Manufacturing expense The sum of the raw material, labor, utilities, maintenance, depreciation, local taxes, etc., expenses. It is the sum of the direct and indirect (fixed and variable) manufacturing expenses.

Marginal cost The incremental cost of making one additional unit without additional investment in facilities

Market capitalization The product of the number of shares of common stock outstanding and the share price.

Market value added A certain future economic value added for a company. It is the present value of the future economic value (EVA) generated by a company. It is a measure of how much value a firm has created.

Minimum acceptable rate of return (MARR) The level of return on investment, at or above the cost of capital, chosen as acceptable for discounting or cutoff purposes.

Net sales price Gross sales price minus freight adjustments.

Net worth The sum of the stockholders' investment plus surplus, or total assets minus total liabilities.

Nominal interest The number applied loosely to describe the annual interest rate.

Obsolescence The occurrence of decreasing value of physical assets due to technological changes rather than physical deterioration.

Operating expense The sum of the manufacturing expense for a product and the general, administrative, and selling expenses.

Operating margin The gross margin minus the general, administrative, and selling expenses.

Opportunity cost The estimate of values that are forgone by undertaking one alternative instead of another one.

Payout time (payback period) The time to recover the fixed capital investment from profit plus depreciation. It is usually after taxes but not always.

Preferred stock Stock having claims that it commands over common stock, with the preference related to dividends. The holders of such stock receive dividends before any distribution is made to common stockholders. Preferred stockholders usually do not have voting rights as common stockholders do.

Present worth The value at some datum time (present time) of expenditures, costs, profits, etc., according to a predetermined method of computation. It is the current value of cash flow obtained by discounting.

Production rate The amount of product manufactured in a given time period.

Profitability A term generally applied in a broad sense to the economic feasibility of a proposed venture or an ongoing operation. It is generally considered to be related to return on investment.

Rate of return on investment The efficiency ratio relating profit or cash flow to investment.

Replacement A new facility that takes the place of an older facility with no increase in capacity.

Revenue The net sales received from the sale of a product or a service to a customer.

Sales, administration, research, and engineering expenses (SARE) Overhead expenses incurred as a result of maintaining sales offices and administrative offices and the expense of maintaining research and engineering departments. This item is usually expressed as a percentage of annual net sales.

Sales volume The amount of sales expressed in pounds, gallons, tons, cubic feet, etc., per unit of time.

Salvage value The value that can be realized from equipment or other facilities when taken out of service and sold.

Selling expense Salaries and commissions paid to sales personnel.

Simple interest The interest charges in any time period that is only charged on the principal.

Sinking fund An accounting procedure computed according to a specified procedure to provide capital to replace an asset.

Surplus The excess of earnings over expenses that is not distributed to stockholders.

Tax credit The amount available to a firm as part of its annual return because of deductible expenses for tax purposes. Examples have been research and development expenses, energy tax credit, etc.

Taxes In a manufacturing cost statement, usually property taxes. In an income statement, usually federal and state income taxes.

Time value of money The expected interest rate that capital should or would earn. Money has value with respect to time.

Total operating investment The fixed capital investment, backup capital, auxiliary capital, utilities and services capital, and working capital.

Utilities and services capital Electrical substations, plant sewers, water distribution facilities, and occasionally the steam plant.

Value added The difference between the raw material expense and the selling price of that product.

Value-improving practices (VIPs) Formal structured practices applied to capital projects to improve profitability ("or value") above that which is attained through the application of proven good engineering and project management practices.

Variable expense Any expense that varies directly with production output.

Working capital In the accounting sense, the current assets minus the current liabilities. It consists of the total amount of money invested in raw materials, supplies, goods in process, product inventories, accounts receivable, and cash minus those liabilities due within 1 year.

Made in United States
Orlando, FL
04 November 2026

90200399R00033